Advances in Experimental Medicine and Biology

Volume 1186

More information about this series at http://www.springer.com/series/5584

Kapil Bharti

Editor

Pluripotent Stem Cells in Eye Disease Therapy

Springer

Editor
Kapil Bharti
National Institutes of Health NEI
Bethesda, MD, USA

ISSN 0065-2598   ISSN 2214-8019   (electronic)
Advances in Experimental Medicine and Biology
ISBN 978-3-030-28473-2   ISBN 978-3-030-28471-8   (eBook)
https://doi.org/10.1007/978-3-030-28471-8

This Springer imprint is published by the registered company Springer Nature Switzerland AG
The registered company address is: Gewerbestrasse 11, 6330 Cham, Switzerland

Preface

Stem cells have an inherent ability to regenerate tissues and organs and because of this property they constitute the core of the field of regenerative medicine. However, stem cells alone are not sufficient to bring a regenerative medicine to patients. They must be combined with other areas of biology including but not limited to tissue engineering, bioengineering, drug discovery, immunology, and surgical approaches. I have focused this book on several of these aspects of regenerative medicine by presenting examples of various interdisciplinary science currently being used for one common goal – regeneration of the entire human body.

This book mainly focuses on examples demonstrating the use of pluripotent stem cells and associated technologies in ocular research. For me, there is no better topic than eye research to provide excellent examples of research progressing in the field of regenerative medicine. The ocular research community has been at the forefront of regenerative medicine research for decades: one of first tissues successfully differentiated from pluripotent stem cells is an eye tissue - the retinal pigment epithelium; the first FDA approved gene therapy is for a monogenic childhood form of retinal degeneration; and the first cell therapy trials conducted using embryonic and induced pluripotent stem cell derived therapies are for age-related macular degeneration, a blinding eye disease. All of this happened by no accident. Various factors have contributed towards the success of regenerative medicine research in the eye: (1) blinding diseases are some of the most prevalent and devastating disorders often afflicting children and depriving patients of one of the most critical senses required to communicate with the outside world. This has led to a major push in the scientific community to develop new treatments for eye diseases; (2) several genes and gene mutations affect the eye without major effects on organismal survival. Because of this several gene mutations have accumulated that affect vision but not the overall patient health. This has sparked scientific curiosity about the role of those genes and mutations and in developing treatments for associated disorders; (3) eye is an easily accessible organ. Because of this several surgical interventions and follow up techniques have been tested and discovered over the years. All of this technology is now helping bring new regenerative medicines to patients; (4) eye is an easily accessible part of the brain, exciting neuroscientists to work on the eye. Because of all these reasons, regenerative medicine of the eye has always gained significant attention. I have planned chapters of this book to highlight some of these factors that have contributed to the success of regenerative medicine in the eye.

I would like the readers of this book, the next generation of scientists and physicians, to appreciate the breath of research that is being conducted around the topic of regenerative medicine and the kind of research that is required to bring successful treatments to patients. Some of my favorites examples include: how patient derived iPS cells are being used to learn more about monogenic diseases but then that knowledge is applied to polygenic diseases that are not easy to study using any other model system; how evolutionary conserved immune responses regulate inflammation in the eye and control immune-response against allogeneic stem cell derived cells; combining bioprinting with stem cell technology to develop complex 3D tissues that provide a native-like environment for both disease modeling and drug testing; and co-evolution of stem cell based therapies and the surgical approaches to bring such stem cell based therapies to patients.

The time is right for such a book, because regenerative medicine is at a crossroads right now where the full potential of stem cells has been realized. Scientists are trying to harness this potential by using a multi-disciplinary approach towards solving the most fundamental problems in biology and to perform the most cutting-edge discoveries in this space. It is critical for the next generation of scientist to understand what approach is currently being adopted to bring the latest regenerative treatments to patients..

Regenerative medicine is a constantly growing field. New discoveries are happening every day. One major limitation of this book, which is perhaps a limitation for every such book, is to stay up to date with the most recent discoveries happening in the field. It took more than a year to complete this book and even though all authors did their best to continually update their chapters, it is highly likely that some of the most recent advances have not been mentioned in this book. But the purpose of this book is not to describe the latest discoveries. My goal with this book is to develop a succinct document that provides a global overview of the field and a reference document for understanding how stem cells can be used to decipher pathways involved in eye diseases and to develop treatments for such eye diseases.

For the successful completion of this work, I would like to thank all the authors who have provided their outstanding chapters. These authors have summarized their own work and the work of others. Therefore, it is only appropriate for me to extend my thanks to all of their labs and to other researchers who have spent countless hours for all the discoveries and inventions that have formed the very basis of this book. I would especially like to thank all the scientists whose vision and hard work gave birth to the field of regenerative medicine and all the associated fields. Most of all, my gratitude goes to patients who altruistically donated samples and their time for research that went into this book.

Last but not least, I would like to thank the editorial staff of Springer publisher for their help and patience (with me) while we tried to complete this work.

Bethesda, MD, USA   Kapil Bharti

Introduction

Developmental biologists have long known about the ability of pluripotent stem cells to differentiate into various cell types. These early observations led to a quest that led to successful culture of mouse embryonic stem (ES) cells and coining of the term ES cells in the year 1981 by Gail Martin. This discovery completely revolutionized the scientific approach towards understanding of gene function. Scientists were able to generate transgenic mice with a gene specific knockout to better investigate gene function, to identify phenotypes associated with specific gene knock out and correlate them with patient phenotype associated with specific diseases. Perhaps even more relevant discovery for human biology was the discovery of human embryonic stem cells by Jamie Thompson in the year 1998. This advancement for the first-time allowed scientists to make human tissues in a dish and to study human developmental biology in vitro. The scientific community also realized the potential of human ES cells to provide replacement tissues as treatments for degenerative diseases. This formed a new field of regenerative medicine with a focus on replacement tissues.

In parallel another scientific endeavor was ongoing to better understand the biology of ES cells. Scientists were trying to convert a somatic human cell into an ES-like cells. This work was originally sparked by a 50 years old observation by Sir John Gurdon, who had demonstrated that the cytoplasm of an egg was sufficient to reprogram the nuclear genome of a somatic cell such that the derived cell could now behave like an ES cell. The work led by Dr. Thompson and Dr. Yamanaka led to the discovery of human induced pluripotent stem (iPS) cells: ES-like cells derived from any somatic cell of the body with the capability to differentiate into any other cell or tissue type of the body. This work was even more exciting for the scientific community because it allowed scientist to develop patient specific cells and tissues in vitro and investigate disease pathogenesis in a dish. Furthermore, it provided a possibility of developing autologous cell therapies that might get around the immune-rejection concerns associated with ES cell derived cell therapies. Discoveries of both ES cells and iPS cells have led to major breakthroughs in different aspects of regenerative medicine. This book focuses on such breakthroughs with highlights of both the disease-in-a-dish and the cell therapy aspect of pluripotent stem cells.

The main emphasis of the book is on the use of stem cells in retina research as remarkable progress has occurred in various aspects of eye research using pluripotent stem cells. The three main retinal cell types that have been successfully derived from both ES cells and iPS cells are the retinal pigment epithelium (RPE), the light sensing photoreceptors, and the retinal ganglion cells. RPE is a monolayer of pigmented epithelial cells, located in the outer retina. It is a polarized tissue with specialized microvilli located towards its apical side and a proteinaceous membrane called the Bruch's membrane towards its basal side. RPE interacts with photoreceptor outer segments via its microvilli and performs several functions to maintain health and integrity of photoreceptors throughout its life. RPE also forms the outer blood retina barrier and regulates nutrient and metabolite flow between the choroidal blood supply on its basal side and photoreceptors on its apical side. RPE dysfunctions lead to choroidal atrophy and photoreceptor cell death leading to vision loss.

Photoreceptors are the main light sensing unit of the retina. There are two main types of photoreceptors: rods and cones. Rods are primarily responsible for dim light vision and cones are responsible for bright light and central vision. Rhodopsin and cone opsins located in the outer segments of these two photoreceptor cell types absorb light photons and transmit those signals to the interneurons of the retina. Through the interneurons, electrical signals are transmitted to retinal ganglion cells (RGCs). RGCs are one of the main neuronal cell types of the retina that carry electrical signals to the visual cortex of the brain where the electrical signal is converted into an image. Different types of RGCs carry signals from different areas of the retina, thus regulating dim light and bright light responses.

Clearly, all three cell types are quint-essential for vision and some of the most prevalent blinding eye disease are associated with degeneration of these three cell types. Successful RPE differentiation from stem cells was achieved a few years before the differentiation of photoreceptors and other cell types of the retina. Overtime researchers have further optimized the RPE differentiation protocols developing RPE monolayer as a functionally validated fully polarized and mature tissue derived from both ES and iPS cells. This has allowed researchers to develop relevant disease models and cell therapies using stem cell derived RPE cells. These advances are reflected in three chapters that are focused on the RPE. Photoreceptor and retinal ganglion cell differentiation initially started in 2D cultures but with seminal discoveries of late Dr. Yoshiki Sasai 3D retinal organoid cultures were also established. Current efforts in the photoreceptor field utilizes both 2D and 3D cultures. However, in neither culture methods researchers have been successful in developing fully polarized photoreceptors that contain opsin protein harboring outer segments or demonstrate light responses similar to what is seen in the native retina. This has limited the use of stem cell derived photoreceptors in disease modeling. But efforts continue in the photoreceptor transplantation field because the thinking is that transplanted photoreceptors will continue to mature when present in the in vivo eye environment. RGC research community has been able to develop neurons that contain several key RGC markers and demonstrate an action potential. This has allowed researchers to simulate RGC diseases using patient specific iPS cells. The importance of these three cell types for vision has long stimulated vision researchers to study their basic developmental biology, to explore disease-inducing pathways, and to develop cell therapies to replace degenerated cells in the eye. This book covers research performed in all these areas on all three cell types. In fact, it goes beyond the work directly performed on these cell types and covers advanced tissue engineering approaches used by stem cell researchers to develop 3D eye tissues containing a capillary network and provides a more native-like environment to study tissue-tissue interaction under healthy and diseased conditions. The book also provides a discussion on surgical approaches that have been developed to transplant cell therapies in the eye. A brief synopsis of all the chapters is provided below.

The first chapter by Dalvi et al presents a "classical" use of iPS cell technology for disease-in-a-dish approach focusing on inherited forms of retinal degenerative diseases. One of the key requirements for developing in vitro disease models is the ability to develop functionally validated RPE cells from iPS cells. Authors discuss various human disease models that have been developed using patient-derived iPS cells and discoveries performed using such human disease models both for better understanding of disease pathogenesis and for discovering potential treatments that can be brought back to patients. Authors also address the potential of stem cells for better understanding of more complex diseases such as age-related macular degeneration.

The second chapter by Greene et al demonstrates an application for stem cell derived wild type cells in studying the fundamental biology of RPE cells. Authors use pluripotent stem cell derived RPE cells to better understand the epithelial phenotype of these cells. Loss of RPE epithelial phenotype associated with eye injuries leads to a condition called proliferative vitreoretinopathy (PVR). Authors present an in vitro PVR model developed using iPS cell derived RPE cells, and also discuss high throughput screens that can be performed using iPS cell derived primary cells.

The third chapter by Ben M'Barek et al covers an important topic of stem cell derived cell therapies. Authors discuss in depth how functionally validated RPE cells are differentiated from ES and iPS cells and used to develop cell therapies for retinal degenerative diseases. Currently, two approaches are being tested to deliver RPE cell therapy in the eye: cells in suspension and cells on a scaffold. Authors discuss differences between these two approaches and how synthetic and natural scaffolds are used to develop an RPE-patch for transplantation. Furthermore, they discuss preclinical efficacy data from animal models used to demonstrate functionality of RPE cell therapies and also discuss preliminary safety data from ongoing human trials using stem cell derived RPE.

The fourth chapter by Kramer et al focuses on photoreceptor-based cell therapies and challenges faced during ES or iPS cell derived photoreceptors transplantation in preclinical animal models. Authors discuss challenges with the integration of transplanted photoreceptors in the host retina and the role played by the host retina microenvironment and the immune system in transplant survival and integration. Authors describe genetic and immune modulatory strategies to overcome the immune response against stem cell derived photoreceptors in an allogeneic host.

The success of this work will help bring stem cell-based treatments to a large number of patients.

The fifth chapter by Ohlemacher et al deals with vision restoration downstream of photoreceptors in retinal ganglion cells. These authors explain how functionally authentic RGCs have been differentiated from ES and iPS cells. They discuss gene expression and electrophysiology readouts used to validate ES or iPS cell derived RGCs. Furthermore, authors review advances in developing optic neuropathy and glaucoma disease models using stem cell derived RGCs, approaches used to discover potential new drugs using stem cell derived RGCs, and an ambitious future challenge to develop a cell therapy to replace degenerated RGCs.

The sixth chapter by Stanzel et al highlights a critical example of associated technologies that must be developed to successfully bring stem cell-based therapies to patients. Here authors demonstrate elegant methods that have been developed to transplant stem cell derived RPE cells in suspension, RPE-monolayer patch, photoreceptors, and retinal sheet grafts. Surgeons are trying such transplants through the front of the eye (going through the vitreous) or the back of the eye (going through the sclera). Authors also discuss various preclinical animal models, their advantages and disadvantages for developing surgical techniques for transplantation in the eye.

The seventh chapter by Boutin et al discusses a next generation technology that combines different stem cell derived cells with bioprinting technology to develop a 3D RPE/choroid tissue. Authors discuss how iPS cell derived endothelial cells when bioprinted on one side of a scaffold are capable of forming a capillary network. This capillary network interacts with the RPE monolayer that is grown on the other side of a scaffold. Similar to the native RPE/choroid, the in vitro 3D choroid tissue depends on the RPE for its survival. Authors demonstrate that this 3D tissue can be used to develop advanced 3D models for complex diseases like AMD. This work shows the power of stem cell technologies when combined with tissue engineering.

The work presented in this book summarizes state-of-the art eye research that is being conducted world-wide using stem cells. The success of this work will lead to improved understanding of human eye development and of diseases that affect the eye. Furthermore, this work will lead to the development of potential therapies to treat these blinding eye diseases.

Contents

1 **Pluripotent Stem Cells to Model Degenerative Retinal Diseases: The RPE Perspective** 1
   Sonal Dalvi, Chad A. Galloway, and Ruchira Singh

2 **Utility of Induced Pluripotent Stem Cell-Derived Retinal Pigment Epithelium for an In Vitro Model of Proliferative Vitreoretinopathy** 33
   Whitney A. Greene, Ramesh R. Kaini, and Heuy-Ching Wang

3 **Developing Cell-Based Therapies for RPE-Associated Degenerative Eye Diseases** 55
   Karim Ben M'Barek, Walter Habeler, Florian Regent, and Christelle Monville

4 **Immunological Considerations for Retinal Stem Cell Therapy** 99
   Joshua Kramer, Kathleen R. Chirco, and Deepak A. Lamba

5 **Advances in the Differentiation of Retinal Ganglion Cells from Human Pluripotent Stem Cells** 121
   Sarah K. Ohlemacher, Kirstin B. Langer, Clarisse M. Fligor, Elyse M. Feder, Michael C. Edler, and Jason S. Meyer

6 **Surgical Approaches for Cell Therapeutics Delivery to the Retinal Pigment Epithelium and Retina** 141
   Boris Stanzel, Marius Ader, Zengping Liu, Juan Amaral, Luis Ignacio Reyes Aguirre, Annekatrin Rickmann, Veluchamy A. Barathi, Gavin S. W. Tan, Andrea Degreif, Sami Al-Nawaiseh, and Peter Szurman

7 **3D Engineering of Ocular Tissues for Disease Modeling and Drug Testing** 171
   M. E. Boutin, C. Hampton, R. Quinn, M. Ferrer, and M. J. Song

Index 195

Contributors

Marius Ader DFG Center for Regenerative Therapies Dresden (CRTD), Technische Universität Dresden, Dresden, Germany

Luis Ignacio Reyes Aguirre DFG Center for Regenerative Therapies Dresden (CRTD), Technische Universität Dresden, Dresden, Germany

Sami Al-Nawaiseh Eye Clinic Sulzbach, Knappschaft Hospital Sulzbach, Sulzbach, Saar, Germany

Juan Amaral Stem Cell and Translational Research Unit, National Eye Institute, National Institutes of Health, Bethesda, MD, USA

Veluchamy A. Barathi Singapore Eye Research Institute, Singapore, Singapore

M. E. Boutin National Center for Advancing Translational Sciences (NCATS), Rockville, MD, USA

Kathleen R. Chirco Department of Ophthalmology, University of California San Francisco, San Francisco, CA, USA

Sonal Dalvi Department of Ophthalmology, Flaum Eye Institute, University of Rochester, Rochester, NY, USA

Department of Biomedical Genetics, University of Rochester, Rochester, NY, USA

Andrea Degreif Fraunhofer Institute for Biomedical Engineering, Sulzbach, Saar, Germany

Michael C. Edler Department of Biology, Indiana University Purdue University Indianapolis, Indianapolis, IN, USA

Department of Medical and Molecular Genetics, Indiana University, Indianapolis, IN, USA

Elyse M. Feder Department of Biology, Indiana University Purdue University Indianapolis, Indianapolis, IN, USA

M. Ferrer National Center for Advancing Translational Sciences (NCATS), Rockville, MD, USA

Clarisse M. Fligor Department of Biology, Indiana University Purdue University Indianapolis, Indianapolis, IN, USA

Chad A. Galloway Department of Ophthalmology, Flaum Eye Institute, University of Rochester, Rochester, NY, USA

Department of Biomedical Genetics, University of Rochester, Rochester, NY, USA

Whitney A. Greene Ocular Trauma Task Area, US Army Institute of Surgical Research, JBSA Fort Sam Houston, Houston, TX, USA

Walter Habeler INSERM U861, I-Stem, AFM, Institute for Stem Cell Therapy and Exploration of Monogenic Diseases, Corbeil-Essonnes, France

UEVE UMR861, Corbeil-Essonnes, France

CECS, Association Française contre les Myopathies, Corbeil-Essonnes, France

C. Hampton National Eye Institute (NEI), Bethesda, MD, USA

Ramesh R. Kaini Ocular Trauma Task Area, US Army Institute of Surgical Research, JBSA Fort Sam Houston, Houston, TX, USA

Joshua Kramer Department of Ophthalmology, University of California San Francisco, San Francisco, CA, USA

Deepak A. Lamba Buck Institute for Research on Aging, Novato, CA, USA

Department of Ophthalmology, University of California San Francisco, San Francisco, CA, USA

Kirstin B. Langer Department of Biology, Indiana University Purdue University Indianapolis, Indianapolis, IN, USA

Zengping Liu Department of Ophthalmology, National University of Singapore, Singapore, Singapore

Karim Ben M'Barek INSERM U861, I-Stem, AFM, Institute for Stem Cell Therapy and Exploration of Monogenic Diseases, Corbeil-Essonnes, France

UEVE UMR861, Corbeil-Essonnes, France

CECS, Association Française contre les Myopathies, Corbeil-Essonnes, France

Jason S. Meyer Department of Biology, Indiana University Purdue University Indianapolis, Indianapolis, IN, USA

Department of Medical and Molecular Genetics, Indiana University, Indianapolis, IN, USA

Stark Neurosciences Research Institute, Indiana University, Indianapolis, IN, USA

Christelle Monville INSERM U861, I-Stem, AFM, Institute for Stem Cell Therapy and Exploration of Monogenic Diseases, Corbeil-Essonnes, France

UEVE UMR861, Corbeil-Essonnes, France

Sarah K. Ohlemacher Department of Biology, Indiana University Purdue University Indianapolis, Indianapolis, IN, USA

R. Quinn National Eye Institute (NEI), Bethesda, MD, USA

Florian Regent INSERM U861, I-Stem, AFM, Institute for Stem Cell Therapy and Exploration of Monogenic Diseases, Corbeil-Essonnes, France

UEVE UMR861, Corbeil-Essonnes, France

Annekatrin Rickmann Eye Clinic Sulzbach, Knappschaft Hospital Sulzbach, Sulzbach, Saar, Germany

Ruchira Singh Department of Ophthalmology, Flaum Eye Institute, University of Rochester, Rochester, NY, USA

Department of Biomedical Genetics, University of Rochester, Rochester, NY, USA

UR Stem Cell and Regenerative Medicine Institute, Rochester, NY, USA

Center for Visual Science, University of Rochester, Rochester, NY, USA

M. J. Song National Center for Advancing Translational Sciences (NCATS), Rockville, MD, USA

National Eye Institute (NEI), Bethesda, MD, USA

Boris Stanzel Eye Clinic Sulzbach, Knappschaft Hospital Sulzbach, Sulzbach, Saar, Germany

Fraunhofer Institute for Biomedical Engineering, Sulzbach, Saar, Germany

Department of Ophthalmology, National University of Singapore, Singapore, Singapore

Peter Szurman Eye Clinic Sulzbach, Knappschaft Hospital Sulzbach, Sulzbach, Saar, Germany

Gavin S. W. Tan Singapore Eye Research Institute, Singapore, Singapore

Heuy-Ching Wang Ocular Trauma Task Area, US Army Institute of Surgical Research, JBSA Fort Sam Houston, Houston, TX, USA

Chapter 1
Pluripotent Stem Cells to Model Degenerative Retinal Diseases: The RPE Perspective

Sonal Dalvi, Chad A. Galloway, and Ruchira Singh

Abstract Pluripotent stem cell technology, including human-induced pluripotent stem cells (hiPSCs) and human embryonic stem cells (hESCs), has provided a suitable platform to investigate molecular and pathological alterations in an individual cell type using patient's own cells. Importantly, hiPSCs/hESCs are amenable to genome editing providing unique access to isogenic controls. Specifically, the ability to introduce disease-causing mutations in control (unaffected) and conversely correct disease-causing mutations in patient-derived hiPSCs has provided a powerful approach to clearly link the disease phenotype with a specific gene mutation. In fact, utilizing hiPSC/hESC and CRISPR technology has provided significant insight into the pathomechanism of several diseases. With regard to the eye, the use of hiPSCs/hESCs to study human retinal diseases is especially relevant to retinal pigment epithelium (RPE)-based disorders. This is because several studies have now consistently shown that hiPSC-RPE in culture displays key physical, gene expression and functional attributes of human RPE in vivo. In this book chapter, we will discuss the current utility, limitations, and plausible future approaches of pluripotent stem cell technology for the study of retinal degenerative diseases. Of note, although we will broadly summarize the significant advances made in modeling and studying several

---

Sonal Dalvi and Chad A. Galloway contributed equally to this work.

S. Dalvi · C. A. Galloway
Department of Ophthalmology, Flaum Eye Institute, University of Rochester, Rochester, NY, USA

Department of Biomedical Genetics, University of Rochester, Rochester, NY, USA

R. Singh (✉)
Department of Ophthalmology, Flaum Eye Institute, University of Rochester, Rochester, NY, USA

Department of Biomedical Genetics, University of Rochester, Rochester, NY, USA

UR Stem Cell and Regenerative Medicine Institute, Rochester, NY, USA

Center for Visual Science, University of Rochester, Rochester, NY, USA
e-mail: Ruchira_Singh@URMC.Rochester.edu

© Springer Nature Switzerland AG 2019
K. Bharti (ed.), *Pluripotent Stem Cells in Eye Disease Therapy*,
Advances in Experimental Medicine and Biology 1186,
https://doi.org/10.1007/978-3-030-28471-8_1

1

retinal diseases utilizing hiPSCs/hESCs, our specific focus will be on the utility of patient-derived hiPSCs for (1) establishment of human cell models and (2) molecular and pharmacological studies on patient-derived cell models of retinal degenerative diseases where RPE cellular defects play a major pathogenic role in disease development and progression.

Keywords Age-related macular degeneration · Choroidal neovascularization · Drusen · Human induced pluripotent stem cell (hiPSC) · hiPSC-based disease modeling · Retinal degenerative diseases · Retinitis pigmentosa · Retinal pigment epithelium

1.1 Overview of Retinal Degenerative Diseases with a Focus on RPE Cell Layer

Retinal degenerative diseases (RDDs) are, as a group, one of the leading causes of irreversible vision loss worldwide. These commonly include retinitis pigmentosa (RP), Leber congenital amaurosis (LCA), Stargardt disease, and age-related macular degeneration (AMD), affecting the central part of retina, the macula, responsible for central vision. Among these, AMD is the most common cause of blindness, affecting elderly individuals, over the age of 55. Apart from aging, epidemiological analyses have identified several genetic and environmental risk factors implicated in the onset and progression of AMD [1, 2]. Mendelian retinal disorders such as heterogenous RP, Stargardt disease, LCA, and Best disease are caused primarily by mutation of a single gene resulting in the absence of or a dysfunctional gene product [3–7]. Although RDDs differ phenotypically, the major retinal cell types primarily affected are photoreceptors, retinal pigment epithelium (RPE), and the underlying choroidal vessels. Furthermore, in a subset of inherited and age-related retinal degenerative diseases (e.g., Best disease, Sorsby's fundus dystrophy, AMD), primary molecular defects in the RPE cell layer in the eye have been implicated in disease development and progression [8–11]. This is not surprising, given that RPE cell health and function is essential for photoreceptor survival and thereby vision maintenance. A noninclusive list of RPE functions includes the absorption of light, exchange of biological materials between the photoreceptors and the choroid, the visual cycle, the processing of shed photoreceptor outer segments, and paracrine cellular communication [12, 13]. The absorption of light, specifically its energy, helps to protect the oxygen-rich retina from photo-oxidation. Light absorption occurs through melanin within the melanosomes and other RPE pigments that filter distinct wavelengths of light, protecting the macula from oxidative stress [14]. The hexagonal RPE forms tight junctions, with the cell layer classified as a tight epithelia, forming a diffusion-impermeable layer. As such, RPE must actively transport biological materials between their apically abutting photoreceptors and their basolateral choroid vasculature. This transport is critical to the balance of water, ions, and pH within the subretinal space. RPE facilitates the uptake of glucose and retinol from the choroidal vasculature to the photoreceptors. The retinoids are the vehicle to transform light energy, through isomerization, to neurological signal giving us vision. Retinol is converted to 11-*cis*-retinal, within the RPE, prior to transport to the photoreceptors where it binds rhodopsin and is integrated into the visual cycle. Photoreceptors lack the isomerases necessary to regenerate 11-*cis*-retinal, which reside in the RPE and transport retinol from the choroidal vasculature, making them indispensable in the visual cycle. The recycling of photoreceptors is another important function of the RPE. Photo-oxidation and free radical generation of the mitochondrial-laden photoreceptors exposes their cellular contents to extensive oxidative damage. Removal of much of this damage is accomplished through outer segment shedding, in which photoreceptor outer segments (POS) containing damaged macromolecules are lost and phagocytosed by the RPE. RPE further degrades the POS contents while recycling retinoids for the visual cycle. Paracrine signaling between the photoreceptors and choroid is a highly coordinated process by the RPE. Among the notable proteins secreted from RPE with physiologic effects relevant to this dialog are pigment epithelium-derived factor (PEDF), vascular endothelial growth factor (VEGF), TIMP metallopeptidase inhibitor 3 (TIMP3), and EGF-containing fibulin extracellular matrix protein 1 (EFEMP1) [9, 12, 13]. PEDF is apically secreted by the RPE and is believed to be neuroprotective as well as anti-angiogenic. In juxtaposition, VEGF is basally secreted and promotes stabilization of the choroidal vasculature in addition to promoting angiogenesis [9, 15]. The proteins TIMP3 and EFEMP1 are involved in regulating extracellular matrix (ECM) turnover [16, 17], crucial for maintenance of the Bruch's membrane, and thereby diffusion and transport of biomolecules across the RPE and choroidal vasculature. Not surprisingly, these proteins have been implicated in macular degenerative diseases and are discussed within mechanisms of genetic macular degenerative diseases. In fact, defects in a number of these RPE cell functions (e.g., phagocytosis and degradation of POS, ECM turnover) have been implicated in both AMD and other RDDs that are caused by mutations in genes regulating these cellular processes (e.g., RP caused by *MERTK* mutations affecting POS phagocytosis [18–21], Doyne honeycomb retinal dystrophy (DHRD) caused by an *EFEMP1* mutation, a gene involved in ECM turnover [22, 23]).

1.1.1 Age-Related Macular Degeneration (AMD)

AMD is the leading cause of vision loss in the western world [24]. By the age of 75, approximately 30% of Americans are affected by the disease [25]. AMD exists in two forms, "dry" and "wet" AMD. Currently there is no treatment for "dry" or geographic atrophy, while "wet" AMD has multiple FDA-approved drugs [26]. Initially both forms of AMD present with a similar pathology are characterized by the appearance of intermediate-sized drusen within the macular region of the retina and thickening of the Bruch's membrane. The Bruch's membrane is a five-layer sandwich of connective tissue consisting of RPE basal lamina, an inner collagenous layer, an elastic layer, an outer collagenous layer, and choriocapillaris basal lamina [9, 27]. The thickening of the Bruch's membrane in AMD consists of fibrous long-spacing collagen, which is geographically located between the RPE and the RPE basal lamina. Drusen are observed as whitish-yellow infiltrates by ophthalmoscopy and are composed of biological materials including protein and lipid or lipoproteinacous material. Within the structure of the retina, drusen reside between the basolateral side of the RPE and the Bruch's membrane; however, unlike the ordered collagenous thickening of the membrane, it appears as more random infiltrates. This positioning is likely critical to the role of drusen and membrane thickening in the development of AMD's pathology, disrupting the functional interaction between the RPE and the vasculature behind the retina. Drusen are characterized as either hard or soft, quantitatively by number and size by their size, <63 or <125 μm, respectively, and qualitatively by border appearance, distinct or fuzzy/diffuse [28]. A greater quantity and presentation of soft drusen was the best predictor of advanced AMD in the Beaver Dam 15-year follow-up study [29]. In the "dry" form of AMD, the appearance of drusen is followed by changes in macular retinal pigment, the atrophy of RPE and ultimately the loss of vision. In contrast, the wet form of AMD involves choroidal neovascularization (CNV) or the proliferation of vasculature from the choroid. While the nature of this event remains ill-defined in AMD progression, stimulation of RPE by complement components C3 and C5, common residents of drusen [30] promoted VEGF secretion from the RPE upon acute exposure in a murine model [30]. This suggests the appearance of drusen in early AMD predisposes the choroid to neovascularization while reinforcing the paracrine signaling balance between the RPE and choroid, here in the promotion of angiogenesis [30, 31]. Genetically an increased susceptibility to AMD has been linked to polymorphisms in genes involved in the complement immune response. The genes implicated include Complement factor H (*CFH*), *C3*, Complement factor I (*CFI*), and Complement Factor B (*CFB*) [32], suggesting a disrupted and/or dysregulated immune response may contribute to the pathology. Polymorphisms within the Age-Related Maculopathy Susceptibility-2 (*ARMS2*) gene are linked to increased susceptibility to AMD [33], though the function of this protein in the disease pathology remains undefined.

While no single mutation is responsible for the development of AMD, there are other forms of late-onset retinal degeneration that do result from single-point mutations. Specifically Sorsby's Fundus Dystrophy (SFD) with a mutation in TIMP3 [34] and Doyne's Honeycomb Retinal Dystrophy/Malattia Leventinesse (DHRD/ML) with a mutation in *EFEMP1* [35], both of which are involved in ECM remodeling, is secreted from the RPE. Likewise, mutations in *C1QTNF5* gene (*CTRP5*), an RPE-secreted protein, which is a constituent of Bruch's membrane leads to late-onset retinal degeneration (L-ORD) [36, 37]. Interestingly these diseases appear to share a significant phenotypic overlap, the development of drusen, integral to progression of the pathology. Despite their inheritance through genetic mutation, these diseases have a relatively late-onset, though earlier than that of AMD, suggesting the development of the pathology is time-dependent despite the immediate expression of a mutated gene product. These characteristics make these diseases plausible surrogates to study the complex time-dependent development of AMD.

1.1.2   Sorsby's Fundus Dystrophy (SFD)

SFD is an autosomal dominant macular degenerative disease, which begins to affect patients between their third and fifth decades of life [38]. The earliest symptom is usually a loss of night vision. Patients may then experience rapid central vision loss followed by peripheral vision loss [39]. The pathology of the disease can be similar to that of wet AMD with neovascularization observed in SFD patients, which may lead to the acute vision loss. Alternatively, a slowed progression with peripheral vision loss parallels the pathophysiology of dry AMD with loss of RPE in distinct areas. Common to both forms is the appearance of peripheral drusen between the RPE and the Bruch's membrane. Unlike AMD where a number of mutations across multiple alleles increase the probability of its development, mutation of a single protein, TIMP3, has been defined as causal in SFD [34]. A number of distinct mutations within TIMP3 have been described to be responsible for the pathology of SFD, most of which cause a missense mutation introducing an additional cysteine residue and all of them clustering around exon 5 of the mRNA message [40]. The mutations lie in the c-terminal region of TIMP3, which has been ascribed the function of inhibition of matrix-metalloproteases (MMPs) [41]. The additional cysteine residues and observed TIMP3 aggregates in SFD patient's led investigators to hypothesize that additional disulfide bridges between TIMP3 molecules may drive the aggregation of protein affecting its interaction and inhibition of MMPs. In agreement, a defining pathologic marker of the disease is a thickening of the ECM adjacent to the Bruch's membrane, suggesting this may indeed affect regulation of ECM remodeling machinery. However, functional analysis of these mutations gave varied results with respect to the ability of the mutated protein to inhibit MMP2 despite the increased propensity of the mutants to form oligomers [16]. It is currently postulated that the oligomers of TIMP3 instead disrupted TIMP3 turnover, leading to their observation of the extreme thickening of the Bruch's membrane. Notably, as mentioned previously, similar to AMD, SFD patients present drusen accumulation between the RPE and the Bruch's membrane preceding the retinal atrophy of the RPE and photoreceptors. Furthermore, TIMP3 protein is a prominent constituent of drusen deposits in both AMD and SFD [10, 42], and rare variants in TIMP3 have recently been linked to AMD development [43].

1.1.3   Doyne's Honeycomb Retinal Dystrophy (DHRD/ML)

DHRD/ML is a macular degenerative disease pathologically identified by the appearance of yellowish-white drusen, RPE atrophy, and neovascularization preceding the loss of vision [44, 45]. Patient vision loss generally has onset in the fourth decade of life. DHRD/ML results from the missense mutation (R345W) in *EFEMP1* gene [35]. EFEMP1 is a secreted protein that has been shown to regulate the activity of multiple MMPs (e.g., MMP-2, MMP-9) involved in ECM turnover [23]. The R345W mutation in *EFEMP1* has been demonstrated to inhibit the proper folding and secretion of the protein [17] and initiating the unfolded protein response (UPR) within RPE [46]. Although the persistent stimulation of the UPR promotes apoptotic cell death, the exact mechanism of how R345W mutation in *EFEMP1* leads to the progression of the disease remains ill-defined. It has however been postulated that *EFEMP1* mutation promotes the DHRD/ML pathology in a dominant negative fashion. The dominant negative action of the mutation is supported by the R345W *EFEMP1* knock-in mice developing pathological markers of the disease, including drusen development at the Bruch's membrane [47], while knock out mice are unaffected with respect to macular health [48]. Recent studies utilizing R345W *EFEMP1* knock-in mice and overexpression of mutant *EFEMP1* in cultured RPE cells, ARPE19, and human fetal RPE have implicated a causal role for C3 activation in the formation of drusen/basal deposits in DHRD, due to impaired ECM turnover [23, 47, 49]. Interestingly, EFEMP1 has also been shown to be a binding partner of TIMP3 [17], the gene causal in SFD, and TIMP3 and EFEMP1 colocalize in the drusen deposits underlying DHRD patients [45]. Furthermore, highlighting a plausible common underlying pathological progression between these distinct diseases; AMD, SFD, and DHRD, similar to TIMP3, EFEMP1 is also found within the drusen deposits underlying AMD patients [22].

1.1.4   Late-Onset Retinal Degeneration

Late-onset retinal degeneration (L-ORD) also sometimes referred to as late-onset macular degeneration (L-ORMD) [50, 51] is a rare autosomal dominant retinal dystrophy primarily affecting the interior segment and retina [52, 53]. A single founder mutation (Ser163Arg) in Complement 1q Tumor Necrosis Factor 5 gene (*C1QTNF5*, previously called *CTRP5*) on chromosome 11 is the causative mutation in L-ORD [36, 37]. *C1QTNF5* is commonly expressed in RPE and ciliary epithelium and is comprised of three domains; a signal peptide at the N-terminus, a short collagen repeat (Gly-X-Y), and a globular complement gC1q domain at the C-terminus and is proposed to function in the trimerization and folding of collagen [36, 51]. Affected individuals display normal visual acuity and fundus examination in stage 1 (ages 0–40); however, some may develop long anterior zonular fibers, iris atrophy, and secondary glaucoma [37, 54, 55]. Patients exhibit disease-like symptoms in stage 2 (fourth to fifth decade of life) exhibiting abnormalities in adaptation from light to dark conditions, perimacular yellow spotting, and midperipheral pigmentation in fundus photographic examination [53, 56]. In stage 3 (sixth decade of life), the patients develop CNV and decline in rod and cone function with sudden loss of visual acuity. At stages 2 and 3 with disease features such as subretinal basal deposits and CNV, L-ORD phenotypically resembles other inherited macular dystrophies including SFD, DHRD, as well as AMD [53]. However, it differs from AMD in terms of inheritance pattern and severity of the extension of sub-RPE deposits and atrophy causing decline of both central and peripheral vision [36, 53]. It varies from SFD and DHRD in terms of disease-causing mutations, later-onset yellow spotting in fundus examination and display of abnormal dark adaptation in L-ORD compared to SFD [36, 53] and variability in the geographic confinement of drusen-like deposits and retinal atrophy in L-ORD compared to DHRD [57, 58]. Despite the differences in the onset and disease severity between L-ORD, SFD, and DHRD, the commonalities in the disease phenotype suggests similar disease pathogenesis among them. In fact, similar to DHRD [46], a study by Shu et al. has implicated endoplasmic reticulum (ER) stress in L-ORD pathology by showing misfolding of mutant *C1QTNF5* and its accumulation in the ER [51].

1.1.5   Best Disease (BD)

BD caused by mutations in Bestrophin-1 gene (*BEST1*) is a childhood-onset inherited dominant form of macular dystrophy characterized by subretinal macular deposition of round or oval yellowish vitelliform lesions [6, 59, 60]. Of note, apart from Best disease, mutations in BEST1 can lead to adult-onset vitelliform macular dystrophy (AVMD) [61], autosomal recessive bestrophinopathy (ARB) [62], autosomal dominant vitreoretinochoroidopathy (ADVIRC) [63], and retinitis pigmentosa 50 (RP50) [64]. Importantly, highlighting a role of RPE dysfunction in BD pathology, BEST1 in the eye, is exclusively expressed within the RPE monolayer. Although the disease mechanism by which mutations in BEST1 lead to BD pathology are still under investigation, several studies have now shown a role of BEST1 in regulating calcium and chloride ions in the RPE cell layer [65–68]. Furthermore, a defect in structural contact between the RPE layer and photoreceptor outer segment (POS) and POS handling [68–70] has been implicated in the pathology of BD. This hypothesis is consistent with the abnormal accumulation of autofluorescent material, lipofuscin, (undigested breakdown products of POS) in the retina, an RPE layer of the affected patient eyes, and consequent photoreceptor degeneration and decline in central vision [6, 71, 72].

1.1.6   Retinitis Pigmentosa

With worldwide prevalence ranging from 1:3000 to 1:7000, RP is the most common hereditary degenerative disorder of the retina. It predominantly affects the photoreceptors leading to rod and cone cell death [73, 74]. Common symptoms include night blindness, decline in electroretinogram responses, gradual loss of peripheral vision subsequently leading to irreversible vision loss. With disease-causing mutations identified in more than 85 genes [75, 76], RP can be inherited in autosomal dominant, autosomal recessive, and X-linked pattern. Apart from photoreceptor-specific cellular defects, disease-causing mutations in RPE-specific genes are also known to contribute to RP development. For example, mutations in the genes involved in visual cycle in the RPE cells have been associated with RP. These include autosomal recessive forms of RP caused by mutations in membrane-type frizzled-related protein (*MFRP*) [77, 78], Mer tyrosine kinase receptor (*MERTK*) [79], and cellular retinaldehyde-binding protein (*CRALBP*) [80]. MFRP is a type II transmembrane domain protein shown to localize apically in RPE microvilli with mutant form leading to defective RPE morphology, cell junctions, and loss of microvilli [81]. In the RPE layer, MERTK regulates the recognition and internalization of POS during phagocytosis, and defective MERTK leads to retinal degeneration via failure of POS phagocytosis [19, 20]. Similarly, CRALBP is present in RPE and Muller cells of the retina and serves as a retinoid carrier involved in the oxidation of 11-*cis*-retinol to 11-*cis*-retinal [82, 83] and is critical for visual cycle regulation and thereby vision. Other RPE-specific visual cycle genes/proteins linked with RP development include lecithin retinol acyltransferase (*LRAT*) and RPE-specific 65 kDa (*RPE65*). Specifically, mutations in *LRAT* and *RPE65* have been reported to account for early-onset forms of RP [84, 85].

1.2   Animal Models of Retinal Degenerative Diseases

The generation of murine models, either knock-in or knockout, has been common practice to the study of genetic diseases. In the study of RDDs, murine models have been generated incorporating known environmental stressors or targeting various genes associated with the disease, with varied results in terms of the replication of disease pathology.

Early murine models of AMD focused on environmental factors contributing to AMD, namely the correlation to obesity and metabolic disease [86, 87]. In these models that were fed a high-fat diet, both age and high-fat diet correlated with an increased thickness of the Bruch's membrane in addition to deposits described as electron lucent particles. However, the thickening of the membrane was not the organized collagenous network observed in aging humans and the particles observed at the Bruch's membrane, and RPE did not contain cholesterol. Efforts to induce hyperlipidemia and elevated cholesterol through gene ablation or transgenic mice have also been used to examine their effect on the development of AMD. The ablation of the apolipoprotein E (*APOE*) gene, a lipid carrier in the blood, resulted in increased thickness of the Bruch's membrane with the appearance of membrane-bound materials [86]. Deletion of the low-density lipoprotein (LDL) receptor again resulted in thickening of the Bruch's membrane with increased lipid deposition at the membrane [88]. Differing results were observed in mice with a mutant form of very low-density lipoprotein (VLDL) receptor gene where an early CNV event occurs in mice as young as 2 weeks [89]. This rapid onset of CNV is seemingly in opposition to CNV in late "wet" AMD.

The role of oxidative stress has been investigated through the deletion of the antioxidant gene including superoxide dismutase 1 (*SOD1*). Along with thickening of the Bruch's membrane, these mice also developed drusen and CNV [31]. These pathological features were apparent only in aged mice, in agreement with the idea that the cumulative insult of oxidative stress promotes AMD. Drusen from these mice contained C5, consistent with human drusen composition [31]. Cigarette smoking, along with being the highest correlative environmental factor with the progression of AMD, promotes oxidative stress. It follows that a smoking mouse model, in which mice were exposed to chronic cigarette smoke, showed increased accumulation of complement factors, C3a, C5, and the membrane attack complex (MAC) C5–9 in the Bruch's membrane relative to control mice [90]. Increases in oxidative stress within the RPE, cellular apoptosis, and thickening of the Bruch's membrane were also observed in response to cigarette smoke exposure in mice [91].

The complement system is well represented within the gene loci associated with the AMD pathology (Sect. 1.1.1). The deposition of C3 and C5 at the Bruch's membrane in the disease pathology also suggests an active role for the complement system in AMD's pathology. It follows that many murine models targeting these genes have been generated to study the disease. Within the retina and RPE/choroid complex of mice, the classical complement factors C1qb, C1r, C1s, C2, and C4 were observed to be constitutively expressed [92, 93]. The alternative pathway components CFH, CFB, C3, and C5 were also detected in these tissues while those in the mannose-binding lectin pathway were extremely low and/or undetectable. The expression of many complement components in the retina and choroid/RPE suggested that murine genetic models of AMD, through manipulation of the complement genes, could plausibly yield mechanistic insight about the disease.

The mutation Y204H within Complement factor H (*CFH*) is associated with a sevenfold increase in the development of AMD in humans [94]. Transgenic mice have been generated that harbor both Y204H point mutation and total gene ablation. The deletion of *CFH* resulted in AMD-like accumulation of drusen, which included C3, and photoreceptor atrophy in aged mice [95]. The accumulation of C3 was also observed in the proper physiologic region, between the basolateral side of the RPE and the Bruch's membrane. The point mutant CFH Y402H mice also displayed an increase in drusen, the thickening of the Bruch's membrane, and C3 accumulation; however, the disease pathology failed to advance to the level of photoreceptor loss [96]. While the knockout of either C3 or C5 proteins has not been reported to describe their effect on the progression of AMD, the reciprocal mice, the knockouts of the C3aR and C5aR receptor proteins have been generated [30]. In the receptor knockout study, the authors hypothesized that C3a and C5a accumulation drives increased VEGF secretion that promotes CNV, given that drusen development and complement factor deposition precede CNV. This hypothesis was based on a prior publication by Ambati et al. in which knockout of Ccr-2 and Ccl2, also known as monocyte chemoattractant protein 1 (MCP1) and its receptor, resulted increased sub-RPE C5a accumulation, by the lack of its removal by the immune system, with a coordinate stimulation of VEGF secretion by the RPE [97]. After initially demonstrating that C5 stimulated VEGF secretion from RPE, laser ablation was used to promote CNV, which was suppressed in the MCP1 receptor knockout mice but not completely abolished [30]. Taken together, these studies support a role for the complement system in the progression of AMD, likely not only in its ability to induce cell death through MAC, but it may also be integral to the promotion of angiogenesis in CNV. This distinction would be relevant to therapeutic intervention against the more damaging form of "wet" AMD. There are currently no murine models of AMD with the manipulation of CFI or CFB, wherein mutations of both are reported to correlate with an increased susceptibility to develop AMD [98, 99].

Using the knowledge of mutations in causative genes identified in inherited late-onset macular dystrophies, murine models for SFD, DHRD, and L-ORD have been developed. In the case of SFD, a knock-in model was created carrying the TIMP3 Ser156Cys mutation [100]. Intriguingly, TIMP3 knock-in mice do not display any pathological manifestations of the human disease. In addition, no formation of basal deposits or ECM thickening was documented in these mice [100]. A knock-in mice model of DHRD with EFEMP1 R345W mutation [47] show increased basal deposits directly linking C3 activation to the formation of basal deposits [49]. A few different murine models of L-ORD [50, 101–103] (heterozygous knock-in mice $Ctrp5^{+/-}$, (hC1QTNF5(S163R)-HA) $C1qtnf5^{+/Ser163Arg}$, $Ctrp5^{+/-};rd^8/rd^8$) have also been established and show a range of key disease-related features of L-ORD and AMD commonly observed in humans [36, 56], including abnormalities in dark adaptation, increased autofluorescent accumulation in the retina, increased abundance in sub-RPE drusen-like deposits, retinal degeneration and abnormalities in Bruch's membrane, and significant loss of photoreceptors cells. In fact, using a recombinant adeno-associated viral (AAV) vector approach, Dinculescu et al. generated a mouse model (hC1QTNF5(S163R)-HA) expressing the $Ctrp5/C1QTNF5$ gene driven by RPE-specific BEST1 promoter to investigate the in vivo consequences of the disease-causing mutation in specific to the RPE [103]. This L-ORD mouse model displayed abnormal accumulation and distribution of the mutant gene within the RPE cells leading to sub-RPE deposits resembling $EFEMP1$ knock-in mice [44, 47]. However, deposits of (hC1QTNF5(S163R)-HA) mice lacked positive staining for lipids, a known component of sub-RPE deposit in L-ORD patients [56, 104] while $C1qtnf5^{+/Ser163Arg}$ mice failed to manifest the L-ORD phenotype throughout its life span [50]. Mouse models both knock-in ($Best1^{+/W93C}$, $Best1^{W93C/W93C}$) and knockout ($Best1^{-/-}$) have also been used to investigate Best disease pathophysiology in vivo [67, 105, 106]. Zhang et al. generated a knock-in mouse carrying the disease-causing mutation W93C in $BEST1$ [67]. This mouse model harbored several of the BD-related features including reduced light peak, lipofuscin accumulation in the RPE, and serous/debris-filled retinal detachment. They also noticed disrupted photoreceptor outer segments suggesting partial impairment of POS phagocytosis by RPE in BD [67]. However, other Best1 knockout mouse models failed to recapitulate the ocular phenotypes of BD. Similarly, contradictory results of functional tests evaluating BEST1 function (Cl⁻ channel, Ca^{2+}-activated Cl⁻ channel (CaCC), volume regulation) were found in the distinct BD mouse models [67, 105, 106]. Several groups have also generated transgenic mice models to investigate the causative role of RPE-specific genes in the pathogenesis of RP. For example, the Royal College of Surgeons (RCS) rats harboring the MERTK mutation associated with early onset of RP [107] are a well-studied model of retinal dystrophy [18, 108]. Retinal degeneration in the RCS rat is associated with defects in POS phagocytosis by the RPE leading to accumulation of phagocytosed OS with alterations in OS length eventually affecting the photoreceptors [108–110]. A similar degenerative retinal phenotype has also been reported in the mer^{kd} mice corroborating the causative role of MERTK mutations in RP [19]. A preclinical mouse model of RP, $Mfrp^{rd6}/Mfrp^{rd6}$ [111, 112], that shows progressive degeneration of the retina and photoreceptors is commonly used to test in vitro RP-related therapies [81, 113]. Of note, although the RP mouse models have provided significant insights into the human disease pathomechanism, contrasting observations have been made in MFRP mutant mice models and MFRP patients [81, 114, 115]. For example, $Mfrp^{rd6}/Mfrp^{rd6}$ and $Mfrp$174delG mice models displayed an increased number of RPE microvilli with no alterations in their length, while the shortened and reduced number of microvilli was reported by Won et al. in the mouse model $Mfrp^{rd6}/Mfrp^{rd6}$ [115]. In contrast, electron microscopy demonstrated loss of apical RPE microvilli in patients with mutant MFRP [81].

Apart from extensively used murine models, other animal models including those derived from rats, pigs, rabbits, and non-human primates have proved to be invaluable in procurement of our current knowledge about the histological features and pathophysiology of specific RDDs. However, apart from non-human primates, the other available animal models display considerable differences with respect to genetic background and physiology within the human retina. For example, a major disadvantage of rodent models is the complete lack of macula, which is the site of disease manifestation, and hence, they fail to recapitulate the AMD disease phenotype as seen in humans. On the other hand, non-human primates anatomically resembling the human retina with the presence of a macula demonstrate early to intermediate features of AMD [116]. However, non-human primate models possess certain obstacles such as difficulty in genetic manipulation, expensive maintenance of non-human primates and slow progression of the disease that does not correlate with that of humans [116].

Other alternatives that have been utilized to interrogate specific RDDs include histopathological examination of the human cadaver eyes and mammalian overexpression systems, including those utilizing immortalized RPE cell lines (e.g., ARPE-19 [117]), and primary RPE cells in culture (e.g., hfRPE and porcine RPE [118–120]). Although these approaches have generated important information about the end-stage pathology and RPE physiology, an optimal platform for understanding the mechanisms behind degenerative diseases of the retina would allow observations of the progression of the disease, i.e., affected cells/tissues from the living human eye progressing through the disease. This is especially relevant, given that postmortem samples are rarely available and are at the end stage of the disease and thus provide no insight into the early events that were causal in the disease development.

1.3 The Pluripotent Stem Cell Technology and Its Utility for Studying Retinal/RPE-Based Disorders

Access to biological samples from human retina and choroid for cellular and molecular studies had not been possible until the advent of the hiPSC technology. The description of reprogramming factors by Yamanaka and Thomson groups [121, 122] made it possible to generate pluripotent stem cells with the ability to differentiate into any of the three germ line lineages and many of their mature cell types. Since then iPSCs derived from patients have been used as a platform to investigate disease pathophysiology and screen for drugs and possible therapeutic approaches. hiPSCs have been generated from a range of sources including fibroblasts, keratinocytes, lymphocytes, core blood cells, adipocytes, and T cells [123–129]. It is noteworthy that several studies have now demonstrated that (1) retinogenesis in a hiPSC-derived model system follows the time course and sequence of retinal development in vivo [130], (2) major cell type(s) of the retina can be consistently differentiated from hiPSCs [127], and (3) hiPSC-derived retinal cells, including RPE, display several important physical and functional attributes akin to their in vivo counterpart. Furthermore, we and others have demonstrated the utility of hiPSCs for studying (1) human retinogenesis and retinal developmental disorders [22, 69, 131] and (2) retinal degenerative diseases like Stargardt disease, BD, glaucoma, and AMD [69, 123, 132–134]. With regard to disease modeling, mechanistic, and pharmacological studies, hiPSC-derived cell models of inherited RDDs, like BD, SFD, DHRD, ADRD, L-ORD, RP, and AMD have clearly established the utility of hiPSC-derived disease models for studying and pharmacologically targeting retinal diseases, including those caused by RPE dysfunction [20, 22, 69, 135–138].

1.3.1 hiPSC Models to Study AMD and Related Retinal Dystrophies (SFD, DHRD, L-ORD)

As mentioned previously, AMD and related retinal dystrophies (SFD, DHRD/ML, and L-ORD) are characterized by formation of lipid-protein-rich basal deposits (drusen), thickening of Bruch's membrane and eventually loss of RPE/photoreceptor layers [9, 139, 140]. Furthermore, a subset of patients with each disorder (AMD, SFD, DHRD/ML, and L-ORD) can develop vision loss due to CNV where in choroidal vasculature grows into the subretinal space [8, 53, 141, 142]. Although the CNV phenotype can be treated in patients (e.g., AMD, SFD [26, 143]), overall the lack of knowledge of molecular and cellular events occurring during the early stages of these disorders, which are causal in disease pathology (e.g., drusen formation), has been detrimental to our ability to develop rational drug therapies.

In a landmark study, Saini et al. utilized hiPSC-RPE derived from AMD patients to (1) investigate the early molecular events in the disease development and (2) test the efficacy of specific drugs in modulating the effected cellular pathway [137]. Specifically, in this study, researchers generated hiPSC-RPE from patients diagnosed with AMD who were homozygous/heterozygous for *ARMS2/HTRA1* and age-matched unaffected controls that showed protective alleles at both loci. Although no differences in baseline RPE characteristics, including expression pattern of RPE signature genes and transepithelial resistance, were found in control vs. AMD hiPSC-RPE, AMD hiPSC-RPE displayed increased expression of drusen components (APOE, amyloid-beta or Aβ) and complement pathway genes [137].

Furthermore, AMD hiPSC-RPE compared to control hiPSC-RPE displayed increased basal secretion of complement protein (C3). Notably, utilizing AMD hiPSC-RPE cultures in long-term experiments spanning 3–12 weeks, the authors demonstrated the ability of a specific drug, nicotinamide (NAM), in suppressing the expression of genes and/or secretion of proteins associated with drusen formation (APOE, APOJ, VEGF-A) and complement pathway (CFH and C3) [137].

In the first study to mimic the drusen phenotype in patient-derived hiPSC-RPE cultures, Galloway et al. utilized hiPSC-RPE from patients with SFD, DHRD, and autosomal dominant radial drusen (ADRD) [22]. Specifically, by utilizing the prolonged culture life of hiPSC-RPE cells and "aging" control (hiPSC-derived from unaffected family members and/or isogenic gene-corrected hiPSC line) and SFD, DHRD, and ADRD hiPSC-RPE cultures (≥90 days in culture), the authors were able to show the presence of basal deposits in both control and patient-derived hiPSC-RPE monocultures. Importantly, basal deposits in patient-derived hiPSC-RPE culture were significantly more than control hiPSC-RPE cultures, present beneath the COL4-positive basement membrane, and demonstrated the presence of several drusen-characteristic proteins, APOE, TIMP3, and EFEMP1. Furthermore, consistent with observed ECM alterations in SFD, DHRD, and ADRD eyes, patient-derived hiPSC-RPE cultures compared to control hiPSC-RPE showed increased accumulation of a specific ECM protein, COL4. Ultimately by using hiPSC-RPE monocultures, from (1) patients with known genetic defect affecting RPE cells (*TIMP3* in SFD; *EFEMP1* in DHRD) and (2) patients with an unidentified genetic defect (ADRD), this study confirmed the causal role of RPE defects in instigating two specific disease hallmarks of AMD and related macular dystrophies, drusen formation, and ECM accumulation.

Gamal et al. utilized a combination of hiPSC-based disease modeling and tissue-on-a-chip approaches to model the events in healthy vs. diseased (L-ORD) RPE following an electrical insult mimicking damage to the RPE cell [144]. Specifically, hiPSC-RPE derived from an affected patient with L-ORD and an unaffected sibling were grown as a monolayer on Electric Cell-Substrate Impedance Sensing (ECIS) microelectrode arrays [145, 146]. The tissue-on-chip approach was then used to investigate the ability of control vs. L-ORD hiPSC-RPE to repair following damage induced by electric wound. Notably, L-ORD hiPSC-RPE demonstrated an impaired rate of wound healing by displaying a reduced rate of migration and dissimilar migration patterns. Of note, for effective cell-substrate attachment and release, a cell line should display optimal migration rate with intermediate adhesion levels [147]. The authors concluded that the reduced migration rates in L-ORD hiPSC-RPE could be accounted for by its stronger adhesion properties to the cell-substrate compared to the control hiPSC-RPE.

Chang et al. generated hiPSCs from the T cells of patients with intermediate and advanced dry AMD and further differentiated them into RPE for use in disease modeling and pharmacological studies [128]. Although, AMD hiPSC-RPE and control hiPSC-RPE showed similar expression of RPE-specific markers such as RLBP1, RPE65, MITF, and PAX6, AMD hiPSC-RPE displayed higher accumulation of endogenous reactive oxygen species (ROS). The increased levels of ROS in AMD hiPSC-RPE cultures were further exacerbated by treatment with H_2O_2. Interestingly, screening of several candidate drugs demonstrated that treatment with curcumin leads to significant reduction in ROS levels in AMD hiPSC-RPE cells. This is an important finding given the fact that oxidative stress has been implicated to cause RPE cell damage in AMD [128].

Yang et al. generated hiPSC-RPE from AMD patients and utilized Bis-retinoid N-retinyl-N-retinylidene ethanolamine (A2E) and blue light exposure to "age" these cells in culture. Interestingly, in comparison to control hiPSC-RPE derived from individuals with homozygous protective haplotype (G–Wt–G; G–Wt–G) for AMD susceptibility, AMD hiPSC-RPE derived from patients with known AMD risk alleles (heterozygous T-in/del-A; G–Wt–G, and homozygous T-in/del-A; T-in/del-A) and showed impaired SOD2 activity accompanied with elevated levels of reactive oxygen species (ROS), thus providing a potential link between oxidative stress and AMD development in individuals harboring the AMD risk alleles (T-in/del-A; G–Wt–G, T-in/del-A; T-in/del-A) [148]. Another AMD-related risk allele identified by genome-wide association study (GWAS) is the complement H factor (*CFH*), and polymorphisms in the *CFH* gene have been strongly linked to AMD pathogenesis via the activation of complement system [149, 150]. Hallam et al. generated hiPSC-RPE from patients harboring the Y402H mutation in the *CFH* gene with varying disease severity. Notably, in the absence of any extrinsic stressors and consistent with AMD disease pathology, patient-derived hiPSC-RPE showed presence of drusen-like deposits that contained known drusen proteins, APOE and C5b-9. Furthermore, the authors reported increased susceptibly to oxidative stress and defective autophagy in AMD hiPSC-RPE cells. In addition, this study also tested the efficacy of treating patient hiPSC-RPE with UV light as a possible treatment therapy. Remarkably, UV light elicited a different response in the low- and high-risk AMD hiPSC-RPE as assessed by the functional and structural characteristics of RPE cells after UV treatment [136]. Also, assessing the role of oxidative stress in AMD, Garcia et al. utilized in vitro modeling of cellular events associated with chronic oxidative stress related to AMD in RPE in both hiPSC and hESC RPE cells [151]. Specifically, chronic exposure to paraquat, activated the NRF2-KEAP1 pathway following induction of specific effectors during the early and late-stage responses, including upregulation of p21, alterations in the microRNA levels (has-miR-146a, has-miR-29a, has-miR-144, has-miR-200a, has-miR-21, has-miR-27b) and identification of Ai-1, an activator with protective role against oxidative stress. Overall, this study successfully illustrated the antioxidant responses and the protective role of the NRF2 pathway in human RPE cells.

To investigate the pathophysiological pathways contributing toward mitochondrial dysfunction in AMD, Golestaneh et al. derived hiPSC-RPE from two AMD patients with abnormal *ARMS2/HTRA1* alleles and one AMD patient with normal *ARMS2/HTRA1* and protective factor B alleles. In accordance with increased susceptibility of AMD hiPSC-RPE to oxidative stress [148], Golestaneh et al. reported similar observations of increased ROS levels and failure to increase SOD2 expression under conditions of oxidative stress in AMD hiPSC-RPE along with ultrastructural damage and dysfunction of mitochondria. Given that peroxisome proliferator-activated receptor-gamma coactivator (*PGC*)-1α is involved in mito-chondrial biogenesis [152] and silent information regulator T1 (*SIRT1*) is a known regulator of *PGC-1α* [153], the authors sought to further gain insight into the under-lying mechanisms responsible for mitochondrial dysfunction in AMD hiPSC-RPE. Notably, AMD hiPSC-RPE displayed reduced expression of PGC-1α and SIRT1 protein levels possibly due to AMPK inactivation, thus implicating the involvement of AMPK/PGC-1α/SIRT1 pathway in AMD pathogenesis.

1.3.2   hiPSC Models of RP

The heterogeneous nature of RP and involvement of both photoreceptor and RPE cells in the disease pathology has made it difficult to identify the impact of disease-causing mutations on individual cell type (RPE vs. photoreceptors) and their conse-quences for disease development in vivo. Furthermore, available animal models of RP do not fully recapitulate the heterogeneous RP phenotype observed in human patients that develops partially due to differences in genetic make-up of affected individuals [154–157]. These limitations make hiPSC technology particularly attractive to study RP as the disease pathomechanism can be interrogated in an indi-vidual cell type (photoreceptor, RPE) using patient's own cells.

Most hiPSC models of RP have typically been developed using a two-dimensional approach by differentiation of patient-derived hiPSCs into either RPE or photorecep-tors depending on the cell types affected by the disease-causing gene. The Takahashi group was one of the first groups to successfully generate multiple patient-derived hiPSC lines from five distinct RP patients carrying mutations in *RP1* (721Lfs722X), *RP9* (H137L), *PRPH2* (W316G), or *RHO* (G188R) genes [138, 158]. Given that these mutations affect photoreceptor cells, hiPSCs in this study were differentiated into photoreceptor cells. Furthermore, electrophysiological and gene expression analysis confirmed the functional and molecular characteristics of hiPSC-photore-ceptors. Further analysis of patient-derived hiPSC photoreceptors showed elevated oxidative stress and ER stress markers with the selective loss of mature rod photore-ceptor cells, whereas cone photoreceptors remained unaffected [138, 158]. Similar observations were made in hiPSC photoreceptors derived from an RP patient carry-ing a different *RHO* mutation (E181K) [159]. To further corroborate causal role of mutant *RHO* gene on development of the RP-disease phenotype in this study, the authors introduced the mutant *RHO* gene harboring E181K mutation in control hiP-SCs, with similar results. Remarkably, the authors also reverted the observed disease phenotype in hiPSC-photoreceptors by correcting the mutation using helper-depen-dent adenoviral vector [159]. In addition, using hiPSC photoreceptors for drug screening studies, Yoshida et al. demonstrated the protective effect of rapamycin, 5-aminoimidazole-4-carboxyamide ribonucleoside (AICAR), Nuclear Quality Assurance-1 (NQDI-1), and salubrinal on rod photoreceptor cell survival [159].

Tucker et al. demonstrated a novel mutation in a newly identified RP gene encod-ing male-germ cell-associated kinase using an array of sequencing techniques and hiPSC-derived retinal cells [160]. Using a similar approach, applying a combination of sequencing and molecular studies on hiPSC-derived retinal precursor cells, Tucker et al. also identified disease-causing mutations in *USH2A* gene and showed that disease-causing *USH2A* variants lead to protein misfolding and ER stress [123].

A similar approach of using patient-specific hiPSC lines has been utilized to study and pharmacologically target the RPE-disease phenotype in RP patients. Schwartz et al. generated hiPSCs from X-linked RP patients carrying the nonsense mutation c.519C>T (p.R120X) in *RP2* gene and differentiated them into RPE cells [161]. The RP2 protein was not detectable in *RP2* R120X patient-derived hiPSC-RPE cells. In conjunction, *RP2* R120X hiPSC-RPE showed defects in Intraflagellar Transport 20 (IFT20) localization, Golgi cohesion, and G protein beta subunit (Gβ1) trafficking. Remarkably, using translational read-through-inducing drugs (TRIDs), the authors partially recovered RP2 protein and consequently reversed the phenotypic abnormalities observed in *R120X* hiPSC-RPE cells [161]. Using a similar approach and utilizing TRIDs on patient-derived hiPSC-RPE from an individual having RP due to the presence of a nonsense variant of *MERTK* gene, Ramsden et al. were also effective in partially restoring the affected function of MERTK, recognition, and internalization of photoreceptor outer segments (POS), in patient-derived hiPSC-derived RPE cells [21].

Li et al. developed an hiPSC-RPE cell model of an autosomal recessive form of RP with mutations in the Membrane Frizzled-Related Protein (*MFRP*) gene that displayed defects in RPE cell pigmentation, morphology, and tight junction formation [81]. Notably, utilizing a gene therapy approach, the authors reversed the disease-specific phenotype in patient-derived hiPSC-RPE cells by AAV8-mediated delivery of wild-type *MFRP*. Furthermore, this study provided novel insights into the role of MFRP in RPE physiology, including (1) modulating actin polymerization and (2) an antagonistic dose-dependent relationship between MFRP and CTRP5 proteins.

1.3.3   hiPSC Models of Other RPE-Related Disorders

Several groups have demonstrated the role of utilizing iPSC-RPE cells to model pathophysiological events in other retinal degenerative disorders including Best Vitelliform Macular dystrophy (BVMD) [69, 135], Gyrate Atrophy [127], and Leber Congenital Amaurosis (LCA) [162].

Given the lack of animal models that recapitulate the BD pathology [105, 106], exclusive localization of BEST1 in a single-cell type, RPE, in the retina [163] involvement of numerous *BEST1* mutations (>200) [62, 164] in the disease, and phenotypic variability between the several different classified bestrophinopathies (AVMD [61], ARB [62], ADVIRC [63], RP50 [64]), hiPSC-based disease modeling and molecular studies are particularly well-suited for studying BEST1 function and the consequence of specific disease-causing mutations on BEST1/RPE cell function in the disease. Moshfegh et al. generated patient-specific hiPSC-RPE from three different patients harboring the mutations R218H, A243T and L234P in the *BEST1* gene and utilized a novel biosensor imaging system to demonstrate impaired Cl⁻ ion efflux in patient-derived hiPSC-RPE compared to control hiPSC-RPE thus suggesting a putative role of BEST1 in regulation of Cl⁻ ions across RPE cell membrane [66]. Li et al. utilized electrophysiological studies on hiPSC-RPE from patients with two distinct *BEST1* mutations, P274R and I201T, to show defective Ca^{2+} dependent Cl⁻ currents in mutant BEST1 hiPSC-RPE cells [65]. Remarkably, this defect was rescued by viral supplementation of wild-type *BEST1*. Marmorstein et al. used hiPSC disease modeling approach to investigate the pathogenesis of ARB [68]. Their specific focus was on the recessive inheritance pattern of ARB that is postulated to be the result of nonsense mediated decay (NMD) that represents null phenotype for *BEST1* [62]. Utilizing hiPSC-RPE from ARB patients harboring compound heterozygous *BEST1* mutations and unaffected controls, they demonstrated that patient-derived hiPSC-RPE display detectable levels of BEST1 mRNA but reduced levels of mutant *BEST1* compared to control hiPSC-RPE cells. Furthermore, consistent with the disease pathology, ARB hiPSC-RPE in this study also showed impairment of POS internalization and phagocytosis. Highlighting a role of defective POS handling in bestrophinopathies, Singh et al. had also previously utilized hiPSC-based disease modeling on two patients harboring distinct mutations in *BEST1* (A146K and N296H) and showed defects in POS degradation by BD hiPSC-RPE compared to control hiPSC-RPE [69]. Overall, these studies have provided insights into both the function of BEST1 in human RPE cells as well as the pathophysiology underlying bestrophinopathies.

Gyrate Atrophy is a progressive autosomal recessive disorder with childhood-onset inducing diffused atrophy of the choroid, RPE and sensory retina caused by mutations in *OAT1*. Meyer et al. utilized hiPSCs to generate patient-derived optic-vesicle like structures and RPE. Importantly they showed that disease-specific functional defect, reduced OAT activity, could be targeted by both gene repair and Vitamin B6 supplementation of gyrate atrophy-hiPSC-RPE cultures [127].

LCA is a rare autosomal recessive retinal disorder associated with early onset of visual loss, pigmentary and retinal abnormalities, nystagmus and reduced electro-retinogram responses. Mutations in at least 20 different genes, including *RPE65*, have been identified as causative in LCA [165]. Using a combination of genome sequencing and a hiPSC-based approach, Tucker et al. identified a novel intronic *RPE65* mutation, IVS3-11 A>G*RPE65* in LCA. Notably, using an hiPSC-approach the authors demonstrated that the pathogenicity of this novel intronic mutation (IVS3-11 A>G*RPE65*) causes induction of abnormal splicing, translational frame-shift and insertion of a premature stop codon [162].

A recent study by Chichagova et al. utilized a hiPSC-based approach to generate hiPSC-RPE from patients with m.3243A>G mitochondrial DNA mutation that is implicated to manifest a range of neurological and ocular phenotypic features [166]. In relation to ocular disease manifestation, patients harboring the m.3243A>G mutation exhibit progressive vision loss with retinal and macular dystrophy [167–169]. The authors demonstrated RPE dysfunction including inability of patient hiPSC-RPE to efficiently phagocytose POS, correlating with lipofuscin accumulation in postmortem eyes of patients [166, 170, 171]. Additionally, m.3243A>G mutation manifested ultrastructural aberrations of mitochondria, hollowed melanosomes, and decline in apical microvilli abundance in patient hiPSC-RPE cells.

1.4 Innovation, Limitations and Plausible Future Approaches for hiPSC-Based Disease Modeling Strategies Focused on Retinal Degenerative Diseases

The brief synopsis of the hiPSC/hESC studies for interrogating retinal degenerative diseases (Sect. 1.3) shows that this human cell model platform has already been successfully utilized to gain important insights into the pathomechanism of both early onset (e.g., Best disease [65, 66, 68, 69]) and late-onset (e.g., AMD [22, 136, 137, 144, 148, 172]) retinal diseases. Remarkably, the utility of hiPSC-derived target cells as a platform to investigate and therapeutically target RDDs has incorporated a variety of different approaches, including gene therapy and drug screening/testing (Fig. 1.1a). Furthermore, the utility of standalone photoreceptor and RPE cultures (derived from patient's own cells) and when relevant, non-diseased physiological stressor (e.g., POS, serum, A2E) has provided a unique strategy to dissect the singular effect of a specific cell type (photoreceptor vs. RPE), in the absence of complex RPE-photoreceptor interaction, on disease-specific molecular and pathological changes (Fig. 1.1b). For instance, by utilizing patient-derived hiPSC-RPE cells from patients with AMD [136, 137, 144, 148, 172] and related macular dystrophies (SFD, DHRD [22]), several groups have recently shown that cellular defects localized to RPE cells are singularly sufficient to cause both molecular (e.g., alterations in complement pathway genes [22, 136, 137]) and pathological structural alterations (e.g., formation of drusen-like basal deposits [22, 137]) in these diseases. Importantly, these studies have provided a cell culture platform where a precise molecular defect in a specific retinal cell type can be directly linked to disease-characteristic clinical phenotype (e.g., autofluorescence accumulation, drusen formation) in a patient-derived human model of the disease. This is particularly relevant given the fact that numerous RDDs affecting the outer retina, impact both the RPE and photoreceptor cell layer. Furthermore, because the photoreceptor-RPE layer in the retina acts as a functional unit, determining the consequence of cell-specific defects in the disease development and progression in vivo has proven difficult. Of note, the capability to mimic pathological phenotype(s) like autofluorescence accumulation, that is a result of chronic physiological insults and develop over time has been assisted by the fact that hiPSC-derived target cells like RPE and photoreceptors (unlike previous cell culture models) can be cultured for an extended period of time (>3 months) [22, 173–176].

Although major advances have already been made utilizing hiPSC-based disease modeling of specific RDDs, it is important to realize that there are limitations of both using a cell culture model derived from hiPSC/hESCs and current disease

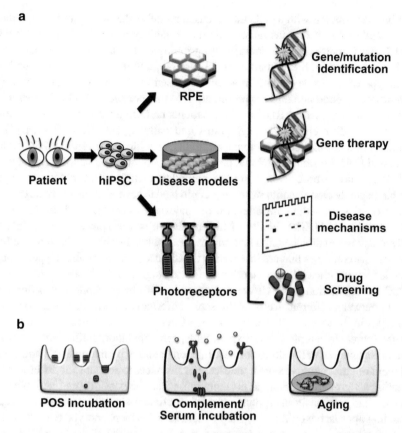

Fig. 1.1 The current hiPSC-based in vitro approaches for retinal degenerative diseases. (**a**) Schematic showing the differentiation of patient-derived hiPSCs to photoreceptor and RPE cells to create a human cell model of the disease that is subsequently utilized for multiple applications including gene/mutation identification in the disease, gene therapy, disease mechanism studies, and drug screening/identification. (**b**) Physiological stressors utilized to metabolically stress hiPSC-RPE in culture that includes exposure to POS, complement/serum, and aging the cells with prolonged time in culture

modeling approaches utilizing a single-cell layer (e.g., photoreceptor, RPE cells) for molecular and therapeutic (gene therapy, drug screening) studies. For instance, hiPSC-RPE in culture lacks the complexity of functional and structural interactions with other cell type(s), including photoreceptors in the retina. Furthermore, hiPSC generation resets their developmental clock, and therefore, hiPSC-RPE in culture are relatively young, and this can pose challenges for studying late-onset diseases, like AMD [177]. As mentioned previously, to overcome these issues, several approaches have been utilized. For example, pharmacological and physiological stressors have been used to metabolically stress and mimic aging of RPE cells in culture [69, 166, 178]. However, there are legitimate concerns with the use of hiPSC/hESC-derived cells that need to be considered in each individual study.

The biggest concern with a cell culture-based model is that of variability, and there are several different reasons underlying the variability in a patient-derived hiPSC model. For instance, a study incorporating multiple patient samples to model and study a disease in an hiPSC-based model system could result in a different cellular phenotype of the same disease in an hiPSC-based model due to variability in the genetic background and clinical presentation of distinct patients. This is consistent with studies in numerous RDDs, where patients harboring the same gene mutation present with different clinical symptoms and pathological characteristics [179]. Another confounding factor is clonal variability, specifically different hiPSC clones generated from the same individual having drastically different cellular characteristics [180]. Furthermore, even in studies limited to a single hiPSC clone per patient sample (a single clone), serial passaging could lead to several changes in the cellular characteristics, including introduction of undesired mutations and chromosomal abnormalities [181]. The fidelity of differentiation is a unique variable in hiPSC-derived cell populations, which becomes relevant when hiPSCs are used to produce two or more cell types that are involved in the disease process, where the percentage of cells forming one cell type vs. other could vary significantly between two distinct hiPSC differentiation runs. Ultimately, there is also the possibility of differences arising between different wells of the same differentiation or differences between cell types in the same well. The resolution of many of these issues is careful study design, increased sample size of experiments, cell population purification (to establish consistency of cell culture for use in downstream experimentation), and establishment of strict quality control metrics. For instance, to account for differences in genetic background and clinical presentation, when the possibility of variable phenotypes has previously been demonstrated in clinical studies, a plausible disease modeling approach would be to group patient samples by phenotype with the inclusion of isogenic control/gene-corrected line (in inherited diseases with known genetic defect) for each patient line. Another approach to resolve this situation would be to use hiPSC-based disease modeling to study diseases that are solely by a singular gene defect with complete penetrance. To account for clonal variability, each individual clone should be thoroughly characterized before experimentation. Furthermore, usage of nonintegrating plasmid vectors for reprogramming and karyotyping of all clones can ensure the absence of any unwanted genetic mutation and chromosomal rearrangement. Genome sequencing can also be used to verify that clones from the same individual are genetically consistent. Most importantly to overcome the variability and likelihood of false-positive results, before formulating any conclusions about the molecular/cellular changes between control and patient cells, it would be important to ascertain that the results are consistent after analysis of multiple clones of patient and control hiPSCs. Similarly, limiting the passage number of hiPSCs utilized in the study helps to maintain genomic integrity that can be monitored through genome sequencing and karyotypic analyses at different passages eliminating variability arising from serial passaging in hiPSC-derived cultures. Ultimately, the baseline characterization of the target cell type (e.g., RPE) in each differentiation run, utilizing some defined criteria (e.g., morphology, pigmentation, polarity) is critical for meaningful experimentation and data interpretation in an hiPSC-derived model system.

The various strategies that have been used for hiPSC-based disease modeling include utilizing isolated cell type(s), complex cell models, 3D culture system, and human–animal chimeras [174, 182, 183]. Of note, as shown by the various examples cited here (see Sect. 1.3), some of these approaches have already been utilized for study of RDDs using hiPSCs. The utility of a single approach is dependent on what phenotypically mimics the disease most accurately with limited complexity. Individual cell types are most relevant when the disease is caused due to dysfunction of a singular cell type, and in case of a genetic disease, the gene responsible for the disease is expressed by that cell type. Alternatively and more complex, interaction between two or more cell type(s) or a specific tissue in its entirety is required for studying the disease pathogenesis and disease pathology. To address such a scenario, complex cell models incorporating multiple cell layers (e.g., photoreceptor-RPE-choriocapillaris) may be necessary (Fig. 1.2). This would first necessitate bioengineering individual cell layers that physiologically and functionally recapitulate their in vivo counterpart (photoreceptor, RPE, choriocapillaris). Important in the proposed scenario, significant advances have already been made into address this goal in both photoreceptors and the RPE [154, 183]. 3D culture systems are also uniquely suited for studying diseases where the complex microenvironment surrounding the cells in vivo is relevant for disease development. 3D culture systems can achieve compartmentalization of different cell type(s) or help promote cellular polarity in an in vitro model system. Finally, the most complex strategy is the generation of humanized animal model. Both hiPSC and hiPSC-differentiated

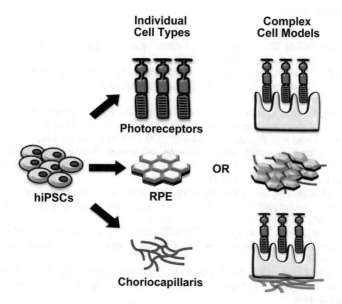

Fig. 1.2 The incorporation of individual cell layers vs. complex cell models in hiPSC-based disease modeling of retinal diseases. Schematic representing the current and future approaches to the utility of a single-cell layer (RPE vs. photoreceptor vs. choriocapillaris) and complex retinal cell models (photoreceptor-RPE, RPE-choriocapillaris, and photoreceptor-RPE-choriocapillaris) in retinal degenerative disease modeling and therapeutic studies (gene therapy, drug screening)

target cells have been injected into animal models (brain, retina), and preliminary studies have shown that these cells can integrate into the mouse tissue and yield humanized neurons and glia [184–186]. However, their utility in modeling and studying RDDs has not yet been established.

1.5   Conclusion

The use of pluripotent stem cell technology has revolutionized our approach to study and therapeutically targeting human diseases and has put the concept of personalized medicine within grasp. This is particularly relevant to RDDs that are a major cause of blindness in children and adults and where few therapies currently exist to target these debilitating disorders. Importantly, over the last decade, we have identified several hundreds of genes that are responsible for specific RDDs (266, https://sph.uth.edu/retnet/sum-dis.htm#B-diseases), but the disease mechanisms in most of these diseases still remain unresolved. Future studies in a patient-derived hiPSC model system are bound to increase our understanding of the molecular basis of several of these diseases thereby leading to the development of rational drug therapies.

References

1. Chen Y, Bedell M, Zhang K (2010) Age-related macular degeneration: genetic and environmental factors of disease. Mol Interv 10:271–281
2. Wang W, Gawlik K, Lopez J et al (2016) Genetic and environmental factors strongly influence risk, severity and progression of age-related macular degeneration. Signal Transduct Target Ther 1:16016
3. Ferrari S, Di Iorio E, Barbaro V, Ponzin D, Sorrentino FS, Parmeggiani F (2011) Retinitis pigmentosa: genes and disease mechanisms. Curr Genomics 12:238–249
4. Tanna P, Strauss RW, Fujinami K, Michaelides M (2017) Stargardt disease: clinical features, molecular genetics, animal models and therapeutic options. Br J Ophthalmol 101:25–30
5. Cremers FPM, van den Hurk JAJM, den Hollander AI (2002) Molecular genetics of Leber congenital amaurosis. Hum Mol Genet 11:1169–1176
6. Boon CJ, Klevering BJ, Leroy BP, Hoyng CB, Keunen JE, den Hollander AI (2009) The spectrum of ocular phenotypes caused by mutations in the BEST1 gene. Prog Retin Eye Res 28:187–205
7. Hartzell HC, Qu Z, Yu K, Xiao Q, Chien LT (2008) Molecular physiology of bestrophins: multifunctional membrane proteins linked to best disease and other retinopathies. Physiol Rev 88:639–672
8. Davis MD, Gangnon RE, Lee LY et al (2005) The Age-Related Eye Disease Study severity scale for age-related macular degeneration: AREDS Report No. 17. Arch Ophthalmol 123:1484–1498
9. Bhutto I, Lutty G (2012) Understanding age-related macular degeneration (AMD): relationships between the photoreceptor/retinal pigment epithelium/Bruch's membrane/choriocapillaris complex. Mol Asp Med 33:295–317

10. Fariss RN, Apte SS, Luthert PJ, Bird AC, Milam AH (1998) Accumulation of tissue inhibitor of metalloproteinases-3 in human eyes with Sorsby's fundus dystrophy or retinitis pigmentosa. Br J Ophthalmol 82:1329–1334
11. Weingeist TA, Kobrin JL, Watzke RC (1982) Histopathology of Best's macular dystrophy. Arch Ophthalmol 100:1108–1114
12. Strauss O (2005) The retinal pigment epithelium in visual function. Physiol Rev 85:845–881
13. Sparrow JR, Hicks D, Hamel CP (2010) The retinal pigment epithelium in health and disease. Curr Mol Med 10:802–823
14. Różanowski B, Burke JM, Boulton ME, Sarna T, Różanowska M (2008) Human RPE melanosomes protect from photosensitized and iron-mediated oxidation but become pro-oxidant in the presence of iron upon photodegradation. Invest Ophthalmol Vis Sci 49:2838–2847
15. Bouck N (2002) PEDF: anti-angiogenic guardian of ocular function. Trends Mol Med 8:330–334
16. Langton KP, McKie N, Smith BM, Brown NJ, Barker MD (2005) Sorsby's fundus dystrophy mutations impair turnover of TIMP-3 by retinal pigment epithelial cells. Hum Mol Genet 14:3579–3586
17. Klenotic PA, Munier FL, Marmorstein LY, Anand-Apte B (2004) Tissue inhibitor of metalloproteinases-3 (TIMP-3) is a binding partner of epithelial growth factor-containing fibulin-like extracellular matrix protein 1 (EFEMP1) - implications for macular degenerations. J Biol Chem 279:30469–30473
18. D'Cruz PM, Yasumura D, Weir J et al (2000) Mutation of the receptor tyrosine kinase gene Mertk in the retinal dystrophic RCS rat. Hum Mol Genet 9:645–651
19. Duncan JL, LaVail MM, Yasumura D et al (2003) An RCS-like retinal dystrophy phenotype in mer knockout mice. Invest Ophthalmol Vis Sci 44:826–838
20. Lukovic D, Artero Castro A, Delgado ABG et al (2015) Human iPSC derived disease model of MERTK-associated retinitis pigmentosa. Sci Rep 5:12910
21. Ramsden CM, Nommiste B, R Lane A et al (2017) Rescue of the MERTK phagocytic defect in a human iPSC disease model using translational read-through inducing drugs. Sci Rep 7:51
22. Galloway CA, Dalvi S, Hung SSC et al (2017) Drusen in patient-derived hiPSC-RPE models of macular dystrophies. Proc Natl Acad Sci U S A 114:E8214–E8223
23. Fernandez-Godino R, Bujakowska KM, Pierce EA (2018) Changes in extracellular matrix cause RPE cells to make basal deposits and activate the alternative complement pathway. Hum Mol Genet 27:147–159
24. Wong WL, Su X, Li X et al (2014) Global prevalence of age-related macular degeneration and disease burden projection for 2020 and 2040: a systematic review and meta-analysis. Lancet Glob Health 2:e106–e116
25. Klein R, Chou CF, Klein BE, Zhang X, Meuer SM, Saaddine JB (2011) Prevalence of age-related macular degeneration in the US population. Arch Ophthalmol 129:75–80
26. Miller JW (2013) Age-related macular degeneration revisited—piecing the puzzle: the LXIX Edward Jackson Memorial Lecture. Am J Ophthalmol 155:1–35.e13
27. Curcio C, Johnson M (2012) Structure, function, and pathology of Bruch's membrane. In: Retina, vol 1, 5th edn. Elsevier, Amsterdam, pp 465–481
28. Spaide RF, Curcio CA (2010) Drusen characterization with multimodal imaging. Retina 30:1441–1454
29. Klein R, Klein BE, Knudtson MD, Meuer SM, Swift M, Gangnon RE (2007) Fifteen-year cumulative incidence of age-related macular degeneration: the Beaver Dam Eye Study. Ophthalmology 114:253–262
30. Nozaki M, Raisler BJ, Sakurai E et al (2006) Drusen complement components C3a and C5a promote choroidal neovascularization. Proc Natl Acad Sci U S A 103:2328–2333
31. Imamura Y, Noda S, Hashizume K et al (2006) Drusen, choroidal neovascularization, and retinal pigment epithelium dysfunction in SOD1-deficient mice: a model of age-related macular degeneration. Proc Natl Acad Sci U S A 103:11282–11287

32. Black JR, Clark SJ (2016) Age-related macular degeneration: genome-wide association studies to translation. Genet Med 18:283–289

33. Sobrin L, Reynolds R, Yu Y et al (2011) ARMS2/HTRA1 locus can confer differential susceptibility to the advanced subtypes of age-related macular degeneration. Am J Ophthalmol 151:345–352.e343

34. Weber BH, Vogt G, Pruett RC, Stohr H, Felbor U (1994) Mutations in the tissue inhibitor of metalloproteinases-3 (TIMP3) in patients with Sorsby's fundus dystrophy. Nat Genet 8:352–356

35. Stone EM, Lotery AJ, Munier FL et al (1999) A single EFEMP1 mutation associated with both Malattia Leventinese and Doyne honeycomb retinal dystrophy. Nat Genet 22:199–202

36. Hayward C, Shu X, Cideciyan AV et al (2003) Mutation in a short-chain collagen gene, CTRP5, results in extracellular deposit formation in late-onset retinal degeneration: a genetic model for age-related macular degeneration. Hum Mol Genet 12:2657–2667

37. Ayyagari R, Mandal MNA, Karoukis AJ et al (2005) Late-onset macular degeneration and long anterior lens zonules result from a CTRP5 gene mutation. Invest Ophthalmol Vis Sci 46:3363–3371

38. Sorsby A, Mason ME (1949) A fundus dystrophy with unusual features. Br J Ophthalmol 33:67–97

39. Hoskin A, Sehmi K, Bird AC (1981) Sorsby's pseudoinflammatory macular dystrophy. Br J Ophthalmol 65:859–865

40. Christensen DRG, Brown FE, Cree AJ, Ratnayaka JA, Lotery AJ (2017) Sorsby fundus dystrophy—a review of pathology and disease mechanisms. Exp Eye Res 165:35–46

41. Langton KP, Barker MD, McKie N (1998) Localization of the functional domains of human tissue inhibitor of metalloproteinases-3 and the effects of a Sorsby's fundus dystrophy mutation. J Biol Chem 273:16778–16781

42. Crabb JW, Miyagi M, Gu X et al (2002) Drusen proteome analysis: an approach to the etiology of age-related macular degeneration. Proc Natl Acad Sci U S A 99:14682

43. Fritsche LG, Igl W, Bailey JN et al (2016) A large genome-wide association study of age-related macular degeneration highlights contributions of rare and common variants. Nat Genet 48:134–143

44. Marmorstein LY, McLaughlin PJ, Peachey NS, Sasaki T, Marmorstein AD (2007) Formation and progression of sub-retinal pigment epithelium deposits in Efemp1 mutation knock-in mice: a model for the early pathogenic course of macular degeneration. Hum Mol Genet 16:2423–2432

45. Marmorstein LY, Munier FL, Arsenijevic Y et al (2002) Aberrant accumulation of EFEMP1 underlies drusen formation in Malattia Leventinese and age-related macular degeneration. Proc Natl Acad Sci U S A 99:13067–13072

46. Roybal CN, Marmorstein LY, Vander Jagt DL, Abcouwer SF (2005) Aberrant accumulation of fibulin-3 in the endoplasmic reticulum leads to activation of the unfolded protein response and VEGF expression. Invest Ophthalmol Vis Sci 46:3973–3979

47. Fu L, Garland D, Yang Z et al (2007) The R345W mutation in EFEMP1 is pathogenic and causes AMD-like deposits in mice. Hum Mol Genet 16:2411–2422

48. McLaughlin PJ, Bakall B, Choi J et al (2007) Lack of fibulin-3 causes early aging and herniation, but not macular degeneration in mice. Hum Mol Genet 16:3059–3070

49. Fernandez-Godino R, Garland DL, Pierce EA (2015) A local complement response by RPE causes early-stage macular degeneration. Hum Mol Genet 24:5555–5569

50. Shu X, Luhmann UFO, Aleman TS et al (2011) Characterisation of a C1qtnf5 Ser163Arg knock-in mouse model of late-onset retinal macular degeneration. PLoS One 6:e27433

51. Shu X, Tulloch B, Lennon A et al (2006) Disease mechanisms in late-onset retinal macular degeneration associated with mutation in C1QTNF5. Hum Mol Genet 15:1680–1689

52. Jacobson SG, Cideciyan AV, Sumaroka A, Roman AJ, Wright AF (2014) Late-onset retinal degeneration caused by C1QTNF5 mutation: sub-retinal pigment epithelium deposits and visual consequences. JAMA Ophthalmol 132:1252–1255

53. Borooah S, Collins C, Wright A, Dhillon B (2009) Late-onset retinal macular degeneration: clinical insights into an inherited retinal degeneration. Br J Ophthalmol 93:284
54. Ayyagari R, Griesinger IB, Bingham E, Lark KK, Moroi SE, Sieving PA (2000) Autosomal dominant hemorrhagic macular dystrophy not associated with the TIMP3 gene. Arch Ophthalmol 118:85–92
55. Subrayan V, Morris B, Armbrecht AM, Wright AF, Dhillon B (2005) Long anterior lens zonules in late-onset retinal degeneration (L-ORD). Am J Ophthalmol 140:1127–1129
56. Kuntz CA, Jacobson SG, Cideciyan AV et al (1996) Sub-retinal pigment epithelial deposits in a dominant late-onset retinal degeneration. Invest Ophthalmol Vis Sci 37:1772–1782
57. Edwards AO, Klein ML, Berselli CB et al (1998) Malattia leventinese: refinement of the genetic locus and phenotypic variability in autosomal dominant macular drusen. Am J Ophthalmol 126:417–424
58. Evans K, Gregory CY, Wijesuriya SD et al (1997) Assessment of the phenotypic range seen in Doyne honeycomb retinal dystrophy. Arch Ophthalmol 115:904–910
59. Xiao Q, Hartzell HC, Yu K (2010) Bestrophins and retinopathies. Pflugers Archiv 460:559–569
60. Marquardt A, Stohr H, Passmore LA, Kramer F, Rivera A, Weber BH (1998) Mutations in a novel gene, VMD2, encoding a protein of unknown properties cause juvenile-onset vitelliform macular dystrophy (Best's disease). Hum Mol Genet 7:1517–1525
61. Kramer F, White K, Pauleikhoff D et al (2000) Mutations in the VMD2 gene are associated with juvenile-onset vitelliform macular dystrophy (Best disease) and adult vitelliform macular dystrophy but not age-related macular degeneration. Eur J Hum Genet 8:286–292
62. Burgess R, Millar ID, Leroy BP et al (2008) Biallelic mutation of BEST1 causes a distinct retinopathy in humans. Am J Hum Genet 82:19–31
63. Yardley J, Leroy BP, Hart-Holden N et al (2004) Mutations of VMD2 splicing regulators cause nanophthalmos and autosomal dominant vitreoretinochoroidopathy (ADVIRC). Invest Ophthalmol Vis Sci 45:3683–3689
64. Davidson AE, Millar ID, Urquhart JE et al (2009) Missense mutations in a retinal pigment epithelium protein, bestrophin-1, cause retinitis pigmentosa. Am J Hum Genet 85:581–592
65. Li Y, Zhang Y, Xu Y et al (2017) Patient-specific mutations impair BESTROPHIN1's essential role in mediating Ca(2+)-dependent Cl(−) currents in human RPE. elife 6:e29914
66. Moshfegh Y, Velez G, Li Y, Bassuk AG, Mahajan VB, Tsang SH (2016) BESTROPHIN1 mutations cause defective chloride conductance in patient stem cell-derived RPE. Hum Mol Genet 25:2672–2680
67. Zhang Y, Stanton JB, Wu J et al (2010) Suppression of Ca^{2+} signaling in a mouse model of Best disease. Hum Mol Genet 19:1108–1118
68. Marmorstein AD, Johnson AA, Bachman LA et al (2018) Mutant Best1 expression and impaired phagocytosis in an iPSC model of autosomal recessive Bestrophinopathy. Sci Rep 8:4487
69. Singh R, Shen W, Kuai D et al (2013) iPS cell modeling of Best disease: insights into the pathophysiology of an inherited macular degeneration. Hum Mol Genet 22:593–607
70. Bakall B, Radu RA, Stanton JB et al (2007) Enhanced accumulation of A2E in individuals homozygous or heterozygous for mutations in BEST1 (VMD2). Exp Eye Res 85:34–43
71. Nordstrom S, Barkman Y (1977) Hereditary maculardegeneration (HMD) in 246 cases traced to one gene-source in Central Sweden. Hereditas 84:163–176
72. Sparrow JR, Duncker T, Woods R, Delori FC (2016) Quantitative fundus autofluorescence in Best vitelliform macular dystrophy: RPE lipofuscin is not increased in non-lesion areas of retina. Adv Exp Med Biol 854:285–290
73. Rivolta C, Sharon D, DeAngelis MM, Dryja TP (2002) Retinitis pigmentosa and allied diseases: numerous diseases, genes, and inheritance patterns. Hum Mol Genet 11:1219–1227
74. Hamel C (2006) Retinitis pigmentosa. Orphanet J Rare Dis 1:40
75. Daiger SP, Sullivan LS, Bowne SJ (2013) Genes and mutations causing retinitis pigmentosa. Clin Genet 84:132–141
76. Ran X, Cai W-J, Huang X-F et al (2014) 'RetinoGenetics': a comprehensive mutation database for genes related to inherited retinal degeneration. Database 2014:bau047

77. Ayala-Ramirez R, Graue-Wiechers F, Robredo V, Amato-Almanza M, Horta-Diez I, Zenteno JC (2006) A new autosomal recessive syndrome consisting of posterior microphthalmos, retinitis pigmentosa, foveoschisis, and optic disc drusen is caused by a MFRP gene mutation. Mol Vis 12:1483–1489

78. Mukhopadhyay R, Sergouniotis PI, Mackay DS et al (2010) A detailed phenotypic assessment of individuals affected by MFRP-related oculopathy. Mol Vis 16:540–548

79. Ostergaard E, Duno M, Batbayli M, Vilhelmsen K, Rosenberg T (2011) A novel MERTK deletion is a common founder mutation in the Faroe Islands and is responsible for a high proportion of retinitis pigmentosa cases. Mol Vis 17:1485–1492

80. Maw MA, Kennedy B, Knight A et al (1997) Mutation of the gene encoding cellular retinaldehyde-binding protein in autosomal recessive retinitis pigmentosa. Nat Genet 17:198–200

81. Li Y, Wu WH, Hsu CW et al (2014) Gene therapy in patient-specific stem cell lines and a preclinical model of retinitis pigmentosa with membrane frizzled-related protein defects. Mol Ther 22:1688–1697

82. Stecher H, Gelb MH, Saari JC, Palczewski K (1999) Preferential release of 11-cis-retinol from retinal pigment epithelial cells in the presence of cellular retinaldehyde-binding protein. J Biol Chem 274:8577–8585

83. Saari JC, Nawrot M, Kennedy BN et al (2001) Visual cycle impairment in cellular retinaldehyde binding protein (CRALBP) knockout mice results in delayed dark adaptation. Neuron 29:739–748

84. Sweeney MO, McGee TL, Berson EL, Dryja TP (2007) Low prevalence of LRAT mutations in patients with Leber congenital amaurosis and autosomal recessive retinitis pigmentosa. Mol Vis 13:588–593

85. Thompson DA, Gyürüs P, Fleischer LL et al (2000) Genetics and phenotypes of RPE65 mutations in inherited retinal degeneration. Invest Ophthalmol Vis Sci 41:4293–4299

86. Dithmar S, Sharara NA, Curcio CA et al (2001) Murine high-fat diet and laser photochemical model of basal deposits in Bruch membrane. Arch Ophthalmol 119:1643–1649

87. Miceli MV, Newsome DA, Tate DJ Jr, Sarphie TG (2000) Pathologic changes in the retinal pigment epithelium and Bruch's membrane of fat-fed atherogenic mice. Curr Eye Res 20:8–16

88. Rudolf M, Winkler B, Aherrahou Z, Doehring LC, Kaczmarek P, Schmidt-Erfurth U (2005) Increased expression of vascular endothelial growth factor associated with accumulation of lipids in Bruch's membrane of LDL receptor knockout mice. Br J Ophthalmol 89:1627–1630

89. Heckenlively JR, Hawes NL, Friedlander M et al (2003) Mouse model of subretinal neovascularization with choroidal anastomosis. Retina 23:518–522

90. Wang AL, Lukas TJ, Yuan M, Du N, Tso MO, Neufeld AH (2009) Autophagy and exosomes in the aged retinal pigment epithelium: possible relevance to drusen formation and age-related macular degeneration. PLoS One 4:e4160

91. Fujihara M, Nagai N, Sussan TE, Biswal S, Handa JT (2008) Chronic cigarette smoke causes oxidative damage and apoptosis to retinal pigmented epithelial cells in mice. PLoS One 3:e3119

92. Luo C, Chen M, Xu H (2011) Complement gene expression and regulation in mouse retina and retinal pigment epithelium/choroid. Mol Vis 17:1588–1597

93. Luo C, Zhao J, Madden A, Chen M, Xu H (2013) Complement expression in retinal pigment epithelial cells is modulated by activated macrophages. Exp Eye Res 112:93–101

94. Klein RJ, Zeiss C, Chew EY et al (2005) Complement factor H polymorphism in age-related macular degeneration. Science 308:385–389

95. Coffey PJ, Gias C, McDermott CJ et al (2007) Complement factor H deficiency in aged mice causes retinal abnormalities and visual dysfunction. Proc Natl Acad Sci U S A 104:16651–16656

96. Ufret-Vincenty RL, Aredo B, Liu X et al (2010) Transgenic mice expressing variants of complement factor H develop AMD-like retinal findings. Invest Ophthalmol Vis Sci 51:5878–5887

97. Ambati J, Anand A, Fernandez S et al (2003) An animal model of age-related macular degeneration in senescent Ccl-2- or Ccr-2-deficient mice. Nat Med 9:1390–1397
98. Alexander P, Gibson J, Cree AJ, Ennis S, Lotery AJ (2014) Complement factor I and age-related macular degeneration. Mol Vis 20:1253–1257
99. Gold B, Merriam JE, Zernant J et al (2006) Variation in factor B (BF) and complement component 2 (C2) genes is associated with age-related macular degeneration. Nat Genet 38:458–462
100. Weber BH, Lin B, White K et al (2002) A mouse model for Sorsby fundus dystrophy. Invest Ophthalmol Vis Sci 43:2732–2740
101. Chavali VRM, Khan NW, Cukras CA, Bartsch D-U, Jablonski MM, Ayyagari R (2011) A CTRP5 gene S163R mutation knock-in mouse model for late-onset retinal degeneration. Hum Mol Genet 20:2000–2014
102. Sahu B, Chavali VR, Alapati A et al (2015) Presence of rd8 mutation does not alter the ocular phenotype of late-onset retinal degeneration mouse model. Mol Vis 21:273–284
103. Dinculescu A, Min S-H, Dyka FM et al (2015) Pathological effects of mutant C1QTNF5 (S163R) expression in murine retinal pigment epithelium. Invest Ophthalmol Vis Sci 56:6971–6980
104. Milam AH, Curcio CA, Cideciyan AV et al (2000) Dominant late-onset retinal degeneration with regional variation of sub-retinal pigment epithelium deposits, retinal function, and photoreceptor degeneration. Ophthalmology 107:2256–2266
105. Milenkovic A, Brandl C, Milenkovic VM et al (2015) Bestrophin 1 is indispensable for volume regulation in human retinal pigment epithelium cells. Proc Natl Acad Sci U S A 112:E2630
106. Marmorstein LY, Wu J, McLaughlin P et al (2006) The light peak of the electroretinogram is dependent on voltage-gated calcium channels and antagonized by Bestrophin (Best-1). J Gen Physiol 127:577–589
107. Gal A, Li Y, Thompson DA et al (2000) Mutations in MERTK, the human orthologue of the RCS rat retinal dystrophy gene, cause retinitis pigmentosa. Nat Genet 26:270–271
108. Bok D, Hall MO (1971) The role of the pigment epithelium in the etiology of inherited retinal dystrophy in the rat. J Cell Biol 49:664–682
109. Dowling JE, Sidman RL (1962) Inherited retinal dystrophy in the rat. J Cell Biol 14:73–109
110. LaVail MM, Battelle BA (1975) Influence of eye pigmentation and light deprivation on inherited retinal dystrophy in the rat. Exp Eye Res 21:167–192
111. Hawes NL, Chang B, Hageman GS et al (2000) Retinal degeneration 6 (rd6): a new mouse model for human retinitis punctata albescens. Invest Ophthalmol Vis Sci 41:3149–3157
112. Kameya S, Hawes NL, Chang B, Heckenlively JR, Naggert JK, Nishina PM (2002) Mfrp, a gene encoding a frizzled related protein, is mutated in the mouse retinal degeneration 6. Hum Mol Genet 11:1879–1886
113. Velez G, Tsang SH, Tsai YT et al (2017) Gene therapy restores Mfrp and corrects axial eye length. Sci Rep 7:16151
114. Fogerty J, Besharse JC (2011) 174delG mutation in mouse MFRP causes photoreceptor degeneration and RPE atrophy. Invest Ophthalmol Vis Sci 52:7256–7266
115. Won J, Smith RS, Peachey NS et al (2008) Membrane frizzled-related protein is necessary for the normal development and maintenance of photoreceptor outer segments. Vis Neurosci 25:563–574
116. Pennesi ME, Neuringer M, Courtney RJ (2012) Animal models of age related macular degeneration. Mol Asp Med 33:487–509
117. Glotin AL, Debacq-Chainiaux F, Brossas JY et al (2008) Prematurely senescent ARPE-19 cells display features of age-related macular degeneration. Free Radic Biol Med 44:1348–1361
118. Johnson LV, Forest DL, Banna CD et al (2011) Cell culture model that mimics drusen formation and triggers complement activation associated with age-related macular degeneration. Proc Natl Acad Sci U S A 108:18277–18282

119. Pilgrim MG, Lengyel I, Lanzirotti A et al (2017) Subretinal pigment epithelial deposition of Drusen components including hydroxyapatite in a primary cell culture model. Invest Ophthalmol Vis Sci 58:708–719
120. Radu RA, Hu J, Jiang Z, Bok D (2014) Bisretinoid-mediated complement activation on retinal pigment epithelial cells is dependent on complement factor H haplotype. J Biol Chem 289:9113–9120
121. Takahashi K, Yamanaka S (2006) Induction of pluripotent stem cells from mouse embryonic and adult fibroblast cultures by defined factors. Cell 126:663–676
122. Yu J, Vodyanik MA, Smuga-Otto K et al (2007) Induced pluripotent stem cell lines derived from human somatic cells. Science 318:1917–1920
123. Tucker BA, Mullins RF, Streb LM et al (2013) Patient-specific iPSC-derived photoreceptor precursor cells as a means to investigate retinitis pigmentosa. elife 2:e00824
124. Wiley LA, Burnight ER, Songstad AE et al (2015) Patient-specific induced pluripotent stem cells (iPSCs) for the study and treatment of retinal degenerative diseases. Prog Retin Eye Res 44:15–35
125. Chen FK, McLenachan S, Edel M, Da Cruz L, Coffey PJ, Mackey DA (2014) iPS cells for modelling and treatment of retinal diseases. J Clin Med 3:1511–1541
126. Phillips MJ, Wallace KA, Dickerson SJ et al (2012) Blood-derived human iPS cells generate optic vesicle-like structures with the capacity to form retinal laminae and develop synapses. Invest Ophthalmol Vis Sci 53:2007–2019
127. Meyer JS, Howden SE, Wallace KA et al (2011) Optic vesicle-like structures derived from human pluripotent stem cells facilitate a customized approach to retinal disease treatment. Stem Cells 29:1206–1218
128. Chang Y-C, Chang W-C, Hung K-H et al (2014) The generation of induced pluripotent stem cells for macular degeneration as a drug screening platform: identification of curcumin as a protective agent for retinal pigment epithelial cells against oxidative stress. Front Aging Neurosci 6:191
129. Tucker BA, Anfinson KR, Mullins RF, Stone EM, Young MJ (2013) Use of a synthetic xeno-free culture substrate for induced pluripotent stem cell induction and retinal differentiation. Stem Cells Transl Med 2:16–24
130. Meyer JS, Shearer RL, Capowski EE et al (2009) Modeling early retinal development with human embryonic and induced pluripotent stem cells. Proc Natl Acad Sci U S A 106:16698–16703
131. Phillips MJ, Perez ET, Martin JM et al (2014) Modeling human retinal development with patient-specific induced pluripotent stem cells reveals multiple roles for visual system homeobox 2. Stem Cells 32:1480–1492
132. Tucker BA, Mullins RF, Stone EM (2014) Stem cells for investigation and treatment of inherited retinal disease. Hum Mol Genet 23(R1):R9–R16
133. Schwartz SD, Regillo CD, Lam BL et al (2015) Human embryonic stem cell-derived retinal pigment epithelium in patients with age-related macular degeneration and Stargardt's macular dystrophy: follow-up of two open-label phase 1/2 studies. Lancet 385(9967):509–516
134. Ramsden CM, Powner MB, Carr AJ, Smart MJ, da Cruz L, Coffey PJ (2013) Stem cells in retinal regeneration: past, present and future. Development 140:2576–2585
135. Singh R, Kuai D, Guziewicz KE et al (2015) Pharmacological modulation of photoreceptor outer segment degradation in a human iPS cell model of inherited macular degeneration. Mol Ther 23:1700–1711
136. Hallam D, Collin J, Bojic S et al (2017) An induced pluripotent stem cell patient specific model of complement factor H (Y402H) polymorphism displays characteristic features of age-related macular degeneration and indicates a beneficial role for UV light exposure. Stem Cells 35:2305–2320
137. Saini JS, Corneo B, Miller JD et al (2017) Nicotinamide ameliorates disease phenotypes in a human iPSC model of age-related macular degeneration. Cell Stem Cell 20:635–647.e637
138. Jin ZB, Okamoto S, Xiang P, Takahashi M (2012) Integration-free induced pluripotent stem cells derived from retinitis pigmentosa patient for disease modeling. Stem Cells Transl Med 1:503–509

139. Curcio CA, Medeiros NE, Millican CL (1996) Photoreceptor loss in age-related macular degeneration. Invest Ophthalmol Vis Sci 37:1236–1249
140. Dorey CK, Wu G, Ebenstein D, Garsd A, Weiter JJ (1989) Cell loss in the aging retina. Relationship to lipofuscin accumulation and macular degeneration. Invest Ophthalmol Vis Sci 30:1691–1699
141. Holz FG, Haimovici R, Wagner DG, Bird AC (1994) Recurrent choroidal neovascularization after laser photocoagulation in Sorsby's fundus dystrophy. Retina 14:329–334
142. Michaelides M, Jenkins SA, Brantley JMA et al (2006) Maculopathy due to the R345W substitution in fibulin-3: distinct clinical features, disease variability, and extent of retinal dysfunction. Invest Ophthalmol Vis Sci 47:3085–3097
143. Prager F, Michels S, Geitzenauer W, Schmidt-Erfurth U (2007) Choroidal neovascularization secondary to Sorsby fundus dystrophy treated with systemic bevacizumab (Avastin). Acta Ophthalmol Scand 85:904–906
144. Gamal W, Borooah S, Smith S et al (2015) Real-time quantitative monitoring of hiPSC-based model of macular degeneration on electric cell-substrate impedance sensing microelectrodes. Biosens Bioelectron 71:445–455
145. Giaever I, Keese CR (1984) Monitoring fibroblast behavior in tissue-culture with an applied electric-field. Proc Natl Acad Sci U S A 81:3761–3764
146. Giaever I, Keese CR (1991) Micromotion of mammalian-cells measured electrically. Proc Natl Acad Sci U S A 88:7896–7900
147. Gupton SL, Waterman-Storer CM (2006) Spatiotemporal feedback between actomyosin and focal-adhesion systems optimizes rapid cell migration. Cell 125:1361–1374
148. Yang J, Li Y, Chan L et al (2014) Validation of genome-wide association study (GWAS)-identified disease risk alleles with patient-specific stem cell lines. Hum Mol Genet 23:3445–3455
149. Golestaneh N, Chu Y, Cheng SK, Cao H, Poliakov E, Berinstein DM (2016) Repressed SIRT1/PGC-1α pathway and mitochondrial disintegration in iPSC-derived RPE disease model of age-related macular degeneration. J Transl Med. 14(1):344. Published 2016 Dec 20. https://doi.org/10.1186/s12967-016-1101-8
150. Lores-Motta L, Paun CC, Corominas J et al (2018) Genome-wide association study reveals variants in CFH and CFHR4 associated with systemic complement activation: implications in age-related macular degeneration. Ophthalmology 125:1064
151. Garcia TY, Gutierrez M, Reynolds J, Lamba DA (2015) Modeling the dynamic AMD-associated chronic oxidative stress changes in human ESC and iPSC-derived RPE cells. Invest Ophthalmol Vis Sci 56:7480–7488
152. Liang H, Ward WF (2006) PGC-1alpha: a key regulator of energy metabolism. Adv Physiol Educ 30:145–151
153. Canto C, Auwerx J (2009) PGC-1alpha, SIRT1 and AMPK, an energy sensing network that controls energy expenditure. Curr Opin Lipidol 20:98–105
154. Artero Castro A, Lukovic D, Jendelova P, Erceg S (2018) Concise review: human induced pluripotent stem cell models of retinitis pigmentosa. Stem Cells 36:474–481
155. Brunner S, Skosyrski S, Kirschner-Schwabe R et al (2010) Cone versus rod disease in a mutant Rpgr mouse caused by different genetic backgrounds. Invest Ophthalmol Vis Sci 51:1106–1115
156. Kostic C, Arsenijevic Y (2016) Animal modelling for inherited central vision loss. J Pathol 238:300–310
157. Sorrentino FS, Gallenga CE, Bonifazzi C, Perri P (2016) A challenge to the striking genotypic heterogeneity of retinitis pigmentosa: a better understanding of the pathophysiology using the newest genetic strategies. Eye 30:1542
158. Jin Z-B, Okamoto S, Osakada F et al (2011) Modeling retinal degeneration using patient-specific induced pluripotent stem cells. PLoS One 6:e17084
159. Yoshida T, Ozawa Y, Suzuki K et al (2014) The use of induced pluripotent stem cells to reveal pathogenic gene mutations and explore treatments for retinitis pigmentosa. Mol Brain 7:45

160. Tucker BA, Scheetz TE, Mullins RF et al (2011) Exome sequencing and analysis of induced pluripotent stem cells identify the cilia-related gene male germ cell-associated kinase (MAK) as a cause of retinitis pigmentosa. Proc Natl Acad Sci U S A 108:E569–E576

161. Schwarz N, Carr AJ, Lane A et al (2015) Translational read-through of the RP2 Arg120stop mutation in patient iPSC-derived retinal pigment epithelium cells. Hum Mol Genet 24:972–986

162. Tucker BA, Cranston CM, Anfinson KA et al (2015) Using patient-specific induced pluripotent stem cells to interrogate the pathogenicity of a novel retinal pigment epithelium-specific 65 kDa cryptic splice site mutation and confirm eligibility for enrollment into a clinical gene augmentation trial. Transl Res 166:740–749.e741

163. Marmorstein AD, Marmorstein LY, Rayborn M, Wang X, Hollyfield JG, Petrukhin K (2000) Bestrophin, the product of the Best vitelliform macular dystrophy gene (VMD2), localizes to the basolateral plasma membrane of the retinal pigment epithelium. Proc Natl Acad Sci U S A 97:12758–12763

164. Pasquay C, Wang LF, Lorenz B, Preising MN (2015) Bestrophin 1—phenotypes and functional aspects in Bestrophinopathies. Ophthalmic Genet 36:193–212

165. Chacon-Camacho OF, Zenteno JC (2015) Review and update on the molecular basis of Leber congenital amaurosis. World J Clin Cases 3:112–124

166. Chichagova V, Hallam D, Collin J et al (2017) Human iPSC disease modelling reveals functional and structural defects in retinal pigment epithelial cells harbouring the m.3243A > G mitochondrial DNA mutation. Sci Rep 7:12320

167. Phillips PH, Newman NJ (1997) Mitochondrial diseases in pediatric ophthalmology. J AAPOS 1:115–122

168. Michaelides M, Jenkins SA, Bamiou DE et al (2008) Macular dystrophy associated with the A3243G mitochondrial DNA mutation. Distinct retinal and associated features, disease variability, and characterization of asymptomatic family members. Arch Ophthalmol 126:320–328

169. Gorman GS, Taylor RW (2011) Mitochondrial DNA abnormalities in ophthalmological disease. Saudi J Ophthalmol 25:395–404

170. Rummelt V, Folberg R, Ionasescu V, Yi H, Moore KC (1993) Ocular pathology of MELAS syndrome with mitochondrial DNA nucleotide 3243 point mutation. Ophthalmology 100:1757–1766

171. Chang TS, Johns DR, Walker D, de la Cruz Z, Maumence IH, Green WR (1993) Ocular clinicopathologic study of the mitochondrial encephalomyopathy overlap syndromes. Arch Ophthalmol 111:1254–1262

172. Golestaneh N, Chu Y, Cheng SK, Cao H, Poliakov E, Berinstein DM (2016) Repressed SIRT1/PGC-1alpha pathway and mitochondrial disintegration in iPSC-derived RPE disease model of age-related macular degeneration. J Transl Med 14:344

173. Singh R, Phillips MJ, Kuai D et al (2013) Functional analysis of serially expanded human iPS cell-derived RPE cultures. Invest Ophthalmol Vis Sci 54:6767–6778

174. Deng WL, Gao ML, Lei XL et al (2018) Gene correction reverses ciliopathy and photoreceptor loss in iPSC-derived retinal organoids from retinitis pigmentosa patients. Stem Cell Rep 10:2005

175. Barnea-Cramer AO, Wang W, Lu SJ et al (2016) Function of human pluripotent stem cell-derived photoreceptor progenitors in blind mice. Sci Rep 6:29784

176. Wahlin KJ, Maruotti JA, Sripathi SR et al (2017) Photoreceptor outer segment-like structures in long-term 3D retinas from human pluripotent stem cells. Sci Rep 7:766

177. Rando TA, Chang HY (2012) Aging, rejuvenation, and epigenetic reprogramming: resetting the aging clock. Cell 148:46–57

178. Parmar VM, Parmar T, Arai E, Perusek L, Maeda A (2018) A2E-associated cell death and inflammation in retinal pigmented epithelial cells from human induced pluripotent stem cells. Stem Cell Res 27:95–104

179. Edwards MM, Maddox DM, Won J, Naggert JK, Nishina PM (2007) Genetic modifiers that affect phenotypic expression of retinal diseases. In: Tombran-Tink J, Barnstable CJ

(eds) Retinal degenerations: biology, diagnostics, and therapeutics. Humana, Totowa, NJ, pp 237–255

180. Vitale AM, Matigian NA, Ravishankar S et al (2012) Variability in the generation of induced pluripotent stem cells: importance for disease modeling. Stem Cells Transl Med 1:641–650

181. Steinemann D, Gohring G, Schlegelberger B (2013) Genetic instability of modified stem cells - a first step towards malignant transformation? Am J Stem Cells 2:39–51

182. Saha K, Jaenisch R (2009) Technical challenges in using human induced pluripotent stem cells to model disease. Cell Stem Cell 5:584–595

183. Nguyen HV, Li Y, Tsang SH (2015) Patient-specific iPSC-derived RPE for modeling of retinal diseases. J Clin Med 4:567–578

184. Windrem MS, Osipovitch M, Liu Z et al (2017) Human iPSC glial mouse chimeras reveal glial contributions to schizophrenia. Cell Stem Cell 21:195–208.e196

185. Lund RD, Wang S, Klimanskaya I et al (2006) Human embryonic stem cell-derived cells rescue visual function in dystrophic RCS rats. Cloning Stem Cells 8:189–199

186. Goldman SA, Nedergaard M, Windrem MS (2015) Modeling cognition and disease using human glial chimeric mice. Glia 63:1483–1493

Chapter 2
Utility of Induced Pluripotent Stem Cell-Derived Retinal Pigment Epithelium for an In Vitro Model of Proliferative Vitreoretinopathy

Whitney A. Greene, Ramesh R. Kaini, and Heuy-Ching Wang

Abstract The advent of stem cell technology, including the technology to induce pluripotency in somatic cells, and direct differentiation of stem cells into specific somatic cell types, has created an exciting new field of scientific research. Much of the work with pluripotent stem (PS) cells has been focused on the exploration and exploitation of their potential as cells/tissue replacement therapies for personalized medicine. However, PS and stem cell-derived somatic cells are also proving to be valuable tools to study disease pathology and tissue-specific responses to injury. High-throughput drug screening assays using tissue-specific injury models have the potential to identify specific and effective treatments that will promote wound healing. Retinal pigment epithelium (RPE) derived from induced pluripotent stem cells (iPS-RPE) are well characterized cells that exhibit the phenotype and functions of in vivo RPE. In addition to their role as a source of cells to replace damaged or diseased RPE, iPS-RPE provide a robust platform for in vitro drug screening to identify novel therapeutics to promote healing and repair of ocular tissues after injury. Proliferative vitreoretinopathy (PVR) is an abnormal wound healing process that occurs after retinal tears or detachments. In this chapter, the role of iPS-RPE in the development of an in vitro model of PVR is described. Comprehensive analyses of the iPS-RPE response to injury suggests that these cells provide a physiologically relevant tool to investigate the cellular mechanisms of the three phases of PVR pathology: migration, proliferation, and contraction. This in vitro model will provide valuable information regarding cellular wound healing responses specific to RPE and enable the identification of effective therapeutics.

W. A. Greene · R. R. Kaini · H.-C. Wang (✉)
Ocular Trauma Task Area, US Army Institute of Surgical Research,
JBSA Fort Sam Houston, Houston, TX, USA
e-mail: heuy-ching.h.wang.civ@mail.mil

© Springer Nature Switzerland AG 2019
K. Bharti (ed.), *Pluripotent Stem Cells in Eye Disease Therapy*,
Advances in Experimental Medicine and Biology 1186,
https://doi.org/10.1007/978-3-030-28471-8_2

Keywords Pluripotent stem cells · Induced pluripotent stem cells · Retinal pigment epithelium · Retinal pigment epithelium derived from induced pluripotent stem cells · Proliferative vitreoretinopathy · Wound healing

2.1   Introduction

A stem cell is broadly defined as a self-renewing entity that can generate several differentiated somatic cell types [1, 2]. Based on their potency to differentiate along different lineages of development, they are classified as pluripotent stem (PS) cells or multipotent stem cells [1–4]. PS cells have the ability to differentiate into cells of all three germ layers, so they can potentially produce every somatic cell type in the human body [4]. The first PS cells identified were embryonic stem (ES) cells discovered in 1981 by two groups who derived them directly from mouse blastocysts [5, 6]. The inherent properties of PS cells have created an exciting new field of scientific research, focused on the exploration and exploitation of their potential as cells/tissue replacement therapies for personalized medicine [7]. PS cells-derived somatic cells are also currently being used to model human developmental processes [8, 9]. Importantly, the PS cell-derived somatic cells display the phenotype and functions of their corresponding somatic cell types in the human body; therefore they are pivotal in experimental models of disease pathology [8–11]. In addition to models of disease pathology, stem cell-derived somatic cells are proving to be valuable tools to study tissue-specific responses to injury. High-throughput drug screening assays using tissue-specific injury models have the potential to identify specific and effective treatments that will promote wound healing. This chapter is devoted to a discussion of a tissue-specific injury model that uses retinal pigment epithelium (RPE) derived from induced pluripotent stem (iPS) cells (iPS-RPE) to study cellular responses to injury and to identify potential therapeutic molecules that will promote wound healing.

2.2   Stem Cell Sources

Stem cell research was revolutionized when Yamanaka successfully reprogrammed adult somatic cells into iPS cells by introducing the four pluripotency transcription factors (Oct 4, c-Myc, Klf4, and Sox2) into their genome [12, 13]. Since iPS cells are directly derived from adult cells, they circumvent the ethical dilemmas associated with ES cells, which require the destruction of human embryos. iPS-derived products are also autologous in source; therefore they can be transplanted without the risk of immune rejection [14–16]. Most importantly, adult somatic cells from diseased tissues can be reprogrammed back to the pluripotent state and further differentiated into specific somatic cell types. The stem cell-derived somatic cells are then expanded to provide large numbers of cells for the development of physiologically relevant in vitro models to study disease and injury pathology [17, 18].

2.3   Reprogramming Techniques

Within a decade after the first successful reprogramming of fibroblasts into iPS cells, several alternative methods have been described in the literature. Early iPS cells were generated by using retroviral gene delivery systems to introduce the aforementioned Yamanaka factors into adult somatic cells [12, 13]. However, retroviral vectors insert randomly into the host genome and may cause insertional mutagenesis [19, 20]; in addition, the transgenes may reactivate after differentiation of the cells into somatic cell types [21]. Therefore the development of alternative gene delivery methods became critical; subsequently various gene delivery methods have been investigated to identify safer alternatives, including lentiviral, transposon, and bacteriophage gene delivery systems [22–24]. These methods still depend on integration of transgenes into the genome and can cause insertional mutagenesis [25]. Consequently, ongoing research efforts are directed at developing integration-free iPS cells. Several methods have been defined, such as adenovirus [26], Sendai virus [27], minicircle vector [28], and episomal vectors [24]. In these methods, the transgenes are expressed in the host cell without integration into the genome; they will be diluted overtime due to host cell division and therefore require multiple transfections. Direct delivery of synthetic modified mRNA [29] and recombinant proteins of the reprogramming factors [30] have also been successfully applied to generate iPS cells. These alternate reprogramming methods are advantageous for development of xeno-free protocols that are safer for clinical applications; however, reprogramming efficiency is sharply lowered in nonintegration approaches compared to the genetic integration strategies [15, 31].

When considering a reprogramming strategy for generation of iPS cells for in vitro purposes, safety issues are less important than the efficiency and scalability of the application. In general, lentiviral and retroviral vectors are the most efficient gene delivery systems for reprogramming purposes [31, 32]. However, retroviruses require cell division for integration into the genome, so it may not be suitable for all cell types [31]. In contrast, lentiviruses do not require cell division to integrate into the genome and are capable of infecting a wider range of somatic cell types than the retrovirus. Finally, different combinations of transcription factors have been shown to affect reprogramming depending on the cell types; therefore the reprogramming strategy needs to be tailored based on the source of cells for disease modeling [31].

2.4   RPE Derived from iPS Cells

Recent innovations in stem cell technology have led to the development of protocols for the efficient differentiation of specific somatic cells from both human ES cells and iPS cells. Somatic cells derived from either human ES cells and iPS cells offer many advantages over primary cells and tumor-derived transformed cell lines. Primary cell lines typically have reduced capacity for growth and proliferation in vitro, and tumor-derived cells often acquire undesirable characteristics such as drug resistance. In fact, iPS-derived cells have been used to study a multitude of human disorders, including hematopoetic, hepatic, neuorological, endothelial, cardiovascular, and retinal diseases. For the purposes of this review, the remainder of this discussion is focused specifically on iPS-RPE and their utility for in vitro models of injury, wound healing, and drug screening. iPS-RPE has been characterized by our research group and others [33–39]. iPS-RPE expresses RPE-specific genes including LRAT, RPE65, CRALBP, and PEDF. In addition, iPS-RPE exhibits characteristic hexagonal, pigmented morphology and is able to internalize vitamin A and convert it into 11-*cis* retinaldehyde through a series of enzymatic reactions (Figs. 2.1 and 2.2). The ability to process vitamin A indicates that these cells express all the essential enzymes of the visual cycle and are therefore fully functional RPE [37]. The iPS-RPE so closely mimics in vivo RPE that it has been transplanted into

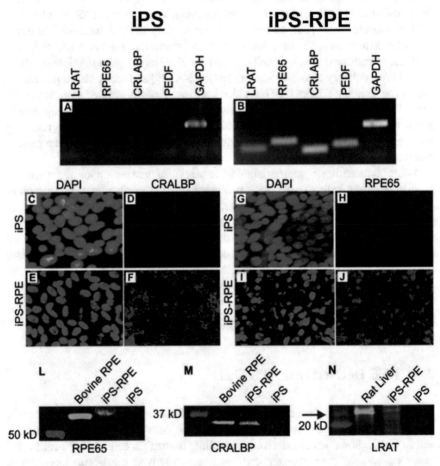

Fig. 2.1 iPS-RPE expresses characteristic RPE genes as indicated by RT-PCR (**a, b**), immunofluorescence (**c–j**), and western blot detection (**l–n**)

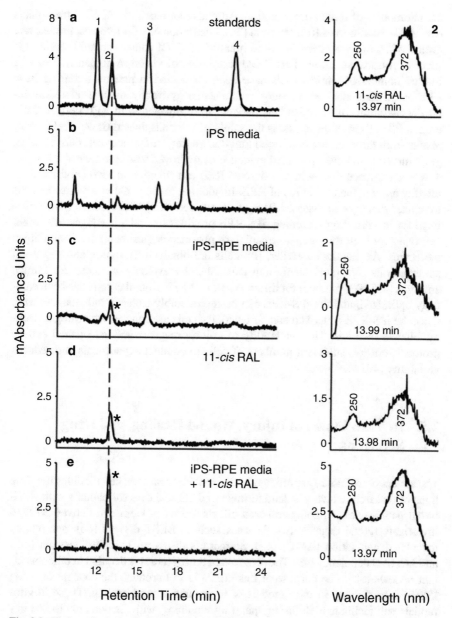

Fig. 2.2 High-performance liquid chromatography (HPLC) analysis of cell culture media demonstrates vitamin A is internalized by iPS-RPE and enzymatically converted to 11-*cis* retinaldehyde (panel **c**)

animal models of retinal degeneration. Animals that received iPS-RPE transplants exhibited restoration of RPE function [34], integration of RPE into host tissues, and even modest improvements in visual function [40]. Of course the gold standard is successful transplantation of RPE and restoration of vision in humans. To date, a limited number of clinical trials have been conducted in patients suffering from RPE-based disorders such as Stargardt's macular dystrophy and age-related macular degeneration. In these studies, patients were injected with RPE derived from human ES cells in order to assess the safety of the transplanted tissue. Preliminary results from those studies confirmed long-term safety of the stem cell-derived RPE, graft survival, and even provided evidence of improved vision in treated eyes [41, 42]. The evidence that stem cell-derived RPE can function in vivo confirms their identity as an authentic source of RPE. In addition to their role as a source of cells to replace damaged or diseased RPE, stem cell-derived RPE provides a robust platform for in vitro drug screening. iPS-RPE provides several advantages over other cell types for high-throughput screening of pharmacological compounds and novel molecules. As mentioned earlier, iPS cells are obtained by reprogramming adult somatic cells into pluripotent stem cells, thereby avoiding the need for human embryos [43]. The protocol for differentiation of RPE from iPS cells has been carefully optimized, and the iPS-RPE can be reproducibly isolated and characterized. Once the iPS-RPE is derived and enriched, the cells are cultured according to well-established methods. The availability of culture reagents and established culture protocols ensures sufficient numbers of cells to conduct experiments and reduces variability [40, 44, 45].

2.5 In Vitro Model of Injury, Wound Healing, and Drug Screening

The advent of technology to efficiently derive RPE from stem cells enables multiple lines of research, which will lead to improved clinical care for patients with RPE-based pathologies. Ongoing research efforts are using stem cell-derived RPE to investigate retinal degenerative diseases such as BEST disease [46], age-related macular degeneration (AMD) [47], Stargardt's disease [41], choroideremia [48], and LCHAD deficiency [49].The progress of those studies is detailed in other excellent reviews [50]. The purpose of this article is to introduce the concept of using iPS-RPE to develop in vitro models of injury and wound healing. These in vitro models will facilitate high-throughput drug screening, with the potential to identify pharmacological compounds that will promote beneficial wound healing after retinal injury. The eye is a highly specialized organ. Although well-protected within the orbital socket, injuries to the eye can have devastating effects on visual function. The delicate tissues of the eye, including the RPE, do not heal well following injury. This abnormal wound healing response leads to the development of fibrosis and scarring. Several post-traumatic intraocular fibrotic conditions have been described, including unilateral subretinal fibrosis [51], aqueous shunt implantation fibrosis

[52], submacular fibrosis [53], occlusion and chronic ocular inflammation [54], traumatic glaucoma [55], traumatic Brown syndrome [56], postsurgical fibrosis after orbital fracture repair [57], post-traumatic orbital reconstruction complications [58–60], commotio retinae causing traumatic maculopathy [61], punctate inner choroidopathy [62], and proliferative vitreoretinopathy (PVR) [63] after perforating injury of the posterior segment [64].

PVR is a frequent secondary outcome of combat-related ocular trauma. A significant number of ocular trauma patients will develop PVR after eye injury; approximately 20% closed-globe and 60% of open-globe eye injuries will progress to develop PVR [65–67]. The risk of PVR after ocular trauma is increased by the presence of two major risk factors: retinal detachment and vitreal hemorrhage. Following trauma to the retina such as a perforating eye injury, PVR is characterized by both the proliferation and migration of cells into the vitreous. The cells secrete extracellular matrix (ECM) proteins; this leads to the formation of either epiretinal or subretinal membranes [66, 68, 69]. The membranes contract over time, pulling on the vitreous and the retina. This in turn causes the secondary retinal detachments that lead to vision loss. In this respect, PVR is considered to be the result of aberrant postinjury wound healing that leads to fibrosis. Currently the standard of care for PVR patients is vitreoretinal surgery to remove these membranes and reattach the retina; although these procedures are anatomically successful, the visual outcomes are poor [70]. Unfortunately, with severe PVR, the retina can subsequently detach again due to development of tractional forces during the postoperative period [68]. As a result, ongoing efforts are focused on the development of therapeutics to inhibit vitreoretinal fibrosis and contractile responses. Several clinical studies have been conducted to compare the results of surgical techniques and medical treatments for PVR. Surgical techniques used include pars plana vitrectomy [71–73], scleral buckle [74–76], inferior retinectomy [77–79], and application of tamponade agents to the inferior retina [80–82]. Medical treatments include intravitreal and oral administration of anti-inflammatory agents such as aspirin [82], triamcinolone [83–85], prednisone [86], dexamethasone [87, 88], chemotherapeutic agents such as daunorubicin [89, 90] and 5-fluorouracil combined with low-molecular weight heparin [91–93], vitamin A derivatives such as isotretinoin [94, 95], anti-angiogenesis agents such as bevacizumab [96], and the antiproliferative agent VIT100 ribozyme [97–99]. In general, the pharmacological compounds tested in clinical studies to mitigate the pathology of PVR fall into five major categories: anti-inflammatory, antiproliferative, anti-angiogenesis, antineoplastic, and vitamin A derivatives. The net result of all these studies remains the same; despite the best efforts of many clinicians and research scientists, a proven therapeutic or prophylactic option for the treatment, repair, or prevention of PVR pathogenesis has not been found [99]. Prevention and successful treatment of PVR remains a challenge for many reasons. PVR progression is the result of a series of complex pathological events and must be considered to be the result of interactions between multiple cell types, extraocular substances, and factors entering the vitreous cavity after the breakdown of the blood–ocular barriers [100]. Trauma that results in separation of the retina from the RPE causes ischemic damage to the retina, inflammation, cell proliferation, and local production of cytokines, chemokines, and growth factors [101, 102]. Breakdown of the blood–ocular barrier is the first step in the pathogenesis of PVR, as this facilitates the infiltration of inflammatory cells from general circulation [103]. Retinal detachment separates the outer nuclear layer from the RPE, which eventually leads to photoreceptor ischemia and death; retinal detachment also activates glial cells in a nonspecific tissue repair response [101, 104]. Upon exposure to the vitreous and extraocular factors, RPE cells transdifferentiate into fibroblastic cells and migrate into the vitreous cavity. Cells in the vitreous cavity produce the characteristic fibrocellular membranes that induce contraction and secondary retinal detachments.

The cellular basis of PVR was first described in 1975 [105–107]. Subsequent studies have implicated pigment epithelial cells (from retina and/or ciliary pigment epithelia) [108], macrophage-like cells, glial cells, and fibroblast-like cells [68, 108] as the essential elements that drive PVR pathogenesis. Polymorphonuclear neutrophils, T and B lymphocytes, platelets, red blood cells, and pigmented cells of unknown origin have also been detected in PVR specimens [63, 66, 68, 109–112]; however, these cells seem to play a secondary role in the pathogenesis of PVR. As researchers continued to search for the primary suspect(s) responsible for the development of PVR, two main cell types came into focus: the Müller glial cells and the RPE. Müller glial cells, activated after retinal injury, are activated in both PVR and retinal detachments that do not develop into PVR. Although Müller cells are important contributors to the pathological process, RPE cells are the key difference between retinal detachment and retinal detachment that progresses into PVR. Several lines of evidence have been developed that support the hypothesis that RPE is critical to the pathogenesis of PVR: (1) RPE cells are present in almost 100% of PVR membranes [100, 108, 113]. (2) The presence of viable RPE in PVR membranes is associated with a poor clinical outcome [114]. (3) RPE cells begin to proliferate within 48 h of retinal detachment [115]. (4) RPE cells are multipotent. Treatment of RPE cells with TGFβ induces conversion from an epithelial phenotype to a fibroblastic phenotype and collagen synthesis [116] and expression of fibronectin, α-smooth muscle actin (α-SMA), and acquisition of myofibroblast characteristics [117]. (5) After transdifferentiation, RPE cells secrete factors and promote contraction of collagen [118], thereby superseding the activity of glial cells in the pathogeneses of PVR [119]. (6) Many animal models of experimental PVR are based on the intravitreal injection of RPE cells to induce the pathology of PVR [120].

The RPE is a single layer of cells positioned at the back of the eye closely juxtaposed to the retina (Fig. 2.3a). Under normal conditions, the RPE comprised highly pigmented hexagonal cells that are nonproliferative when tight cell–cell contacts are maintained, in a state of contact inhibition. However, when a retinal break or detachment disrupts contact inhibition, the RPE at the site of injury undergo epithelial–mesenchymal transition (EMT) and invade the vitreal cavity where they become proliferating, fibroblastic, and contractile cells (Fig. 2.3) [121]. When the blood–retina barrier is compromised due to a perforating eye injury, the RPE is exposed to a variety of cytokines and growth factors in the blood and vitreal fluid. While the exact etiology of PVR remains elusive, the growth factor hypothesis suggests that vitreal growth factors and cytokines are essential drivers of the pathogenesis of this

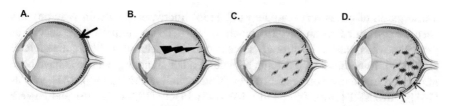

Fig. 2.3 (**a**) The retinal pigment epithelium (RPE) is a single layer of pigmented cells located at the posterior segment of the eye between the retina and choroid (arrow). (**b**) Perforating injuries that cause retinal tears expose RPE to the vitreous cavity. (**c**) RPE will transdifferentiate into fibroblasts due to EMT and migrate into the vitreous cavity. (**d**) Fibroblasts will secrete ECM components to form the epiretinal membrane (yellow fibers). Some fibroblasts will further differentiate into myofibroblasts. The epiretinal membrane will attach to the retina. The myofibroblasts provide the contractile force that causes secondary retinal detachments (arrows)

blinding disease. Studies have identified growth factors/cytokines in the vitreous and the epiretinal/subretinal membranes from PVR patients, including transforming growth factor β (TGFβ), platelet-derived growth factor (PDGF), hepatocyte growth factor (HGF), vascular endothelial growth factor (VEGF), epidermal growth factor (EGF), transforming growth factor α (TGFα), and connective tissue growth factor (CTGF) [122–124]. Among these factors, TGFβ has been proven to be a multifunctional cytokine that stimulates transdifferentiation of RPE into fibroblasts/myofibroblasts and potently induces the expression of contractile proteins, specifically α-SMA [125–129]. Therefore, TGFβ, TGFβ receptor or the TGFβ-dependent pathway (Smad) are promising therapeutic targets for the treatment of PVR. In addition to TGFβ, PDGF expression is strongly associated with PVR pathology. The level of PDGF in the vitreous from patients with PVR is higher than in patients with retinal conditions other than PVR [130, 131]. Similarly, the vitreal levels of PDGFs increase as experimental animals develop PVR [132, 133]. Furthermore, PDGFs are present in PVR membranes, as are activated PDGF receptors [134]. Strategies to inhibit or neutralize the vitreal growth factors/receptors or their cognate signaling pathways may provide effective treatment options.

A number of studies have been conducted in efforts to elucidate the cellular signaling mechanisms that underlie RPE activation in response to injury. In 2002, Hinton et al. examined surgically excised PVR membranes for the presence of HGF and CTGF. Based on the results of that analysis, the authors proposed a model of PVR pathogenesis in which injury to the retina induces an inflammatory response and upregulation of growth factor expression, particularly HGF and CTGF. The RPE separates from the monolayer and forms a multilayered group of dedifferentiated, migratory cells. Exposure to growth factors in the vitreous such as TGFβ causes the dedifferentiated cells to express α-SMA as they convert into myofibroblasts. The myofibroblastic RPE cells contract the extracellular matrix resulting in tractional secondary detachment of the retina [123]. The conversion of RPE into fibroblasts and then myofibroblasts has been determined to be the result of EMT [113, 116, 135]. During EMT, activation of cell surface receptors induces translocation of β-catenin from the cell membrane to the nucleus, where it promotes the transcription of genes that regulate EMT [136]. Increased β-catenin signaling has been demonstrated in an in vitro model of PVR using cultured RPE sheets, as β-catenin was found to be strongly localized in the nuclei of dedifferentiated RPE [137–139]. In addition to β-catenin signaling, other signaling pathways have been implicated in the promotion of EMT of RPE, including Erk1/2 [140, 141], mTOR [141], PI3K/AKT [142], TGFβ [140, 143–145], PDGF [102, 132, 146], and Jagged/ Notch [140], to name a few. Furthermore, a variety of molecules have been shown activate these signaling pathways in RPE, including microRNA-29b [147], thrombin [141, 148–150], reactive oxygen species [151], and of course the growth factors EGF/FGF-2 [152], CTGF [153], HGF [145, 154], PDGF [132, 134], VEGF [102, 155], and TGFβ [145, 156–158]. These research efforts have provided valuable knowledge and insight into the cellular events that contribute to EMT as well as PVR and revealed multiple potential therapeutic targets. In fact, several follow-up studies have examined the ability of pharmaceutical and chemical inhibitors of those pathways to prevent the onset of EMT in RPE cells. DAPT to inhibit Notch [159], C2 ceramide inhibition of CTGF [160], troglitazone [143], glucosamine [161], and genetic inhibitors to target the TGFβ pathway [162, 163], anti-PDGF antibodies [164], trichostatin A to inhibit histone deacetylation [165], the antioxidant n-acetyl-cysteine [166], ROCK inhibitors [158], retinoic acid inhibition of bFGF activity [167], and dasatinib to inhibit tyrosine kinases [168] are a few of the compounds that have been tested in experimental PVR models for their effects on RPE proliferation, migration, and contraction. Although each pathway has specific functions, crosstalk and overlap between pathways potentially contribute to PVR pathogenesis [140], and certainly increase the degree of difficulty in identification of therapeutic targets. With every pathway identified thus far, studies have been attempted in animal models to test the efficacy of pathway inhibition to prevent or repair PVR, with limited success, and as stated earlier, none of these have led to effective treatments for PVR patients. Given the complex nature of PVR pathogenesis, it is not surprising that strategies to hit only one target have yet to yield therapeutic success in the clinic.

In order to develop an in vitro model that accurately recapitulates the pathogenesis of PVR, the cells used for experimental analysis must exhibit the phenotypic and functional characteristics of in vivo RPE. The cells used for these studies should express essential RPE characteristics: pigmentation, hexagonal morphology, expression of RPE-specific genes, and the ability to process vitamin A. Comprehensive analyses of iPS-RPE phenotype and function suggest that these cells provide a physiologically relevant tool to investigate the cellular mechanisms of the three phases of PVR pathology: migration, proliferation, and contraction. As shown in Fig. 2.4, the basis of this in vitro model is a wound-healing assay. The iPS-RPE is grown until fully confluent and pigmented, and then scratched to create a wound (Fig. 2.4a). Studies have shown that the addition of 5% fresh normal vitreous fluid to the iPS-RPE during wound-healing induces dramatically increased expression of α-SMA, indicating a fibrotic response that accurately represents the pathology of PVR (Fig. 2.4b). Analysis of the iPS-RPE during wound healing has demonstrated that the cells undergo EMT at the wound edge, which allows the cells to migrate

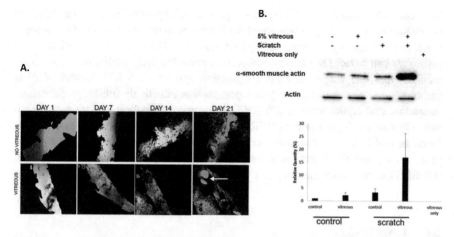

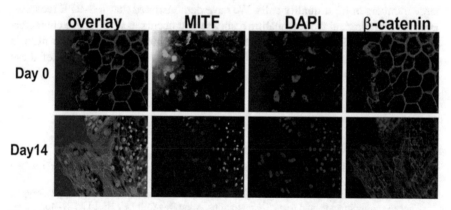

Fig. 2.4 (**a**) Wound healing assay (patent applied for) with iPS-RPE without vitreous exposure (top) and with vitreous exposure (bottom). Note the highly pigmented (black) phenotype of the iPS-RPE. Arrows indicate tears in the monolayer due to contraction. (**b**) Western blot analysis of iPS-RPE demonstrates a significant increase in α-SMA expression after vitreous exposure

Fig. 2.5 Activation of EMT at the wound edge. Immunofluorescence of iPS-RPE during wound healing. β-catenin (red) has translocated from the plasma membrane to the nucleus at the wound edge, indicating activation of EMT. Reduced expression of a RPE marker, microphthalmia-associated transcription factor (MiTF) (green), indicates conversion of the cells to mesenchymal phenotype. Cell nuclei are indicated with DAPI (blue)

into the wound area and proliferate (Fig. 2.5). This model will be used to identify pharmacological agents and novel molecules, such as microRNAs (miRNAs) that inhibit PVR pathogenesis. This model has several advantages over other in vitro models of PVR: (1) The iPS-RPE is of human origin. (2) The iPS-RPE displays the correct phenotype with pigmented, polygonal morphology when cultured to confluence. (3) The iPS-RPE has been thoroughly characterized; the cells express characteristic RPE genes and are able to process vitamin A to complete the visual cycle.

(4) A consistent source of cells ensures experimental reproducibility; the acquisition of reliable data from high-throughput drug screens demands the use of a homogenous, physiologically relevant cell population. (5) The iPS-RPE model of PVR pathology can be used to identify compounds that effectively inhibit the early, intermediate, and late stages of PVR progression. (6) The iPS-RPE model of PVR pathology can be used to identify compounds that effectively inhibit proliferation, migration, and contraction of iPS-RPE during wound healing and vitreous exposure. (7) The iPS-RPE model of PVR pathology can be used to identify compounds (such as miRNA) that promote the early, intermediate, and late stages of PVR progression, and those that activate proliferation, migration, and contraction of iPS-RPE during wound healing and vitreous exposure.

2.6   Conclusion

RPE cells derived from stem cells offer several advantages for the in vitro study of wound healing after retinal injury. They are functionally and phenotypically validated as authentic RPE. They can be expanded under controlled conditions to yield large numbers of high-quality cells. We have demonstrated that iPS-RPE recapitulates the pathological wound healing response that occurs in vivo. These characteristics make iPS-RPE a valuable tool for the study of post-traumatic ocular fibrosis and will enable the identification of novel molecules for the treatment of these blinding disorders.

References

1. Smith AG (2001) Embryo-derived stem cells: of mice and men. Annu Rev Cell Dev Biol 17:435–462
2. Weissman IL, Anderson DJ, Gage F (2001) Stem and progenitor cells: origins, phenotypes, lineage commitments, and transdifferentiations. Annu Rev Cell Dev Biol 17:387–403
3. Gurdon JB, Melton DA (2008) Nuclear reprogramming in cells. Science 322:1811–1815
4. Liao SY, Tse HF (2013) Multipotent (adult) and pluripotent stem cells for heart regeneration: what are the pros and cons? Stem Cell Res Ther 4:151
5. Evans MJ, Kaufman MH (1981) Establishment in culture of pluripotential cells from mouse embryos. Nature 292:154–156
6. Martin GR (1981) Isolation of a pluripotent cell line from early mouse embryos cultured in medium conditioned by teratocarcinoma stem cells. Proc Natl Acad Sci U S A 78: 7634–7638
7. Lancaster MA, Knoblich JA (2014) Organogenesis in a dish: modeling development and disease using organoid technologies. Science 345:1247125
8. Ko HC, Gelb BD (2014) Concise review: drug discovery in the age of the induced pluripotent stem cell. Stem Cells Transl Med 3:500–509
9. Park IH, Arora N, Huo H, Maherali N, Ahfeldt T, Shimamura A, Lensch MW, Cowan C, Hochedlinger K, Daley GQ (2008) Disease-specific induced pluripotent stem cells. Cell 134:877–886

10. Onder TT, Daley GQ (2012) New lessons learned from disease modeling with induced pluripotent stem cells. Curr Opin Genet Dev 22:500–508
11. Jang J, Yoo JE, Lee JA, Lee DR, Kim JY, Huh YJ, Kim DS, Park CY, Hwang DY, Kim HS, Kang HC, Kim DW (2012) Disease-specific induced pluripotent stem cells: a platform for human disease modeling and drug discovery. Exp Mol Med 44:202–213
12. Okita K, Ichisaka T, Yamanaka S (2007) Generation of germline-competent induced pluripotent stem cells. Nature 448:313–317
13. Takahashi K, Yamanaka S (2006) Induction of pluripotent stem cells from mouse embryonic and adult fibroblast cultures by defined factors. Cell 126:663–676
14. Hyun I (2010) The bioethics of stem cell research and therapy. J Clin Invest 120:71–75
15. Kaini RR, Shen-Gunther J, Cleland JM, Greene WA, Wang HC (2016) Recombinant xeno-free vitronectin supports self-renewal and pluripotency in protein-induced pluripotent stem cells. Tissue Eng Part C Methods 22:85. https://doi.org/10.1089/ten.TEC.2015.0180
16. Condic ML, Rao M (2010) Alternative sources of pluripotent stem cells: ethical and scientific issues revisited. Stem Cells Dev 19:1121–1129
17. Cramer AO, MacLaren RE (2013) Translating induced pluripotent stem cells from bench to bedside: application to retinal diseases. Curr Gene Ther 13:139–151
18. Kim K, Zhao R, Doi A, Ng K, Unternaehrer J, Cahan P, Huo H, Loh YH, Aryee MJ, Lensch MW, Li H, Collins JJ, Feinberg AP, Daley GQ (2011) Donor cell type can influence the epigenome and differentiation potential of human induced pluripotent stem cells. Nat Biotechnol 29:1117–1119
19. Oliveira PH, da Silva CL, Cabral JM (2014) Concise review: genomic instability in human stem cells: current status and future challenges. Stem Cells 32:2824–2832
20. Peterson SE, Loring JF (2014) Genomic instability in pluripotent stem cells: implications for clinical applications. J Biol Chem 289:4578–4584
21. Toivonen S, Ojala M, Hyysalo A, Ilmarinen T, Rajala K, Pekkanen-Mattila M, Aanismaa R, Lundin K, Palgi J, Weltner J, Trokovic R, Silvennoinen O, Skottman H, Narkilahti S, Aalto-Setala K, Otonkoski T (2013) Comparative analysis of targeted differentiation of human induced pluripotent stem cells (hiPSCs) and human embryonic stem cells reveals variability associated with incomplete transgene silencing in retrovirally derived hiPSC lines. Stem Cells Transl Med 2:83–93
22. Yu J, Vodyanik MA, Smuga-Otto K, Antosiewicz-Bourget J, Frane JL, Tian S, Nie J, Jonsdottir GA, Ruotti V, Stewart R, Slukvin II, Thomson JA (2007) Induced pluripotent stem cell lines derived from human somatic cells. Science 318:1917–1920
23. Woltjen K, Michael IP, Mohseni P, Desai R, Mileikovsky M, Hamalainen R, Cowling R, Wang W, Liu P, Gertsenstein M, Kaji K, Sung HK, Nagy A (2009) piggyBac transposition reprograms fibroblasts to induced pluripotent stem cells. Nature 458:766–770
24. Ye L, Chang JC, Lin C, Qi Z, Yu J, Kan YW (2010) Generation of induced pluripotent stem cells using site-specific integration with phage integrase. Proc Natl Acad Sci U S A 107:19467–19472
25. Yu J, Hu K, Smuga-Otto K, Tian S, Stewart R, Slukvin II, Thomson JA (2009) Human induced pluripotent stem cells free of vector and transgene sequences. Science 324:797–801
26. Zhou W, Freed CR (2009) Adenoviral gene delivery can reprogram human fibroblasts to induced pluripotent stem cells. Stem Cells 27:2667–2674
27. Fusaki N, Ban H, Nishiyama A, Saeki K, Hasegawa M (2009) Efficient induction of transgene-free human pluripotent stem cells using a vector based on Sendai virus, an RNA virus that does not integrate into the host genome. Proc Jpn Acad Ser B Phys Biol Sci 85:348–362
28. Narsinh KH, Jia F, Robbins RC, Kay MA, Longaker MT, Wu JC (2011) Generation of adult human induced pluripotent stem cells using nonviral minicircle DNA vectors. Nat Protoc 6:78–88
29. Warren L, Manos PD, Ahfeldt T, Loh YH, Li H, Lau F, Ebina W, Mandal PK, Smith ZD, Meissner A, Daley GQ, Brack AS, Collins JJ, Cowan C, Schlaeger TM, Rossi DJ (2010) Highly efficient reprogramming to pluripotency and directed differentiation of human cells with synthetic modified mRNA. Cell Stem Cell 7:618–630

30. Kim D, Kim CH, Moon JI, Chung YG, Chang MY, Han BS, Ko S, Yang E, Cha KY, Lanza R, Kim KS (2009) Generation of human induced pluripotent stem cells by direct delivery of reprogramming proteins. Cell Stem Cell 4:472–476
31. Brouwer M, Zhou H, Nadif KN (2016) Choices for induction of pluripotency: recent developments in human induced pluripotent stem cell reprogramming strategies. Stem Cell Rev 12:54–72
32. Schlaeger TM, Daheron L, Brickler TR, Entwisle S, Chan K, Cianci A, DeVine A, Ettenger A, Fitzgerald K, Godfrey M, Gupta D, McPherson J, Malwadkar P, Gupta M, Bell B, Doi A, Jung N, Li X, Lynes MS, Brookes E, Cherry AB, Demirbas D, Tsankov AM, Zon LI, Rubin LL, Feinberg AP, Meissner A, Cowan CA, Daley GQ (2015) A comparison of non-integrating reprogramming methods. Nat Biotechnol 33:58–63
33. Buchholz DE, Hikita ST, Rowland TJ, Friedrich AM, Hinman CR, Johnson LV, Clegg DO (2009) Derivation of functional retinal pigmented epithelium from induced pluripotent stem cells. Stem Cells 27:2427–2434
34. Carr AJ, Vugler AA, Hikita ST, Lawrence JM, Gias C, Chen LL, Buchholz DE, Ahmado A, Semo M, Smart MJ, Hasan S, da Cruz L, Johnson LV, Clegg DO, Coffey PJ (2009) Protective effects of human iPS-derived retinal pigment epithelium cell transplantation in the retinal dystrophic rat. PLoS One 4:e8152
35. Muniz A, Ramesh KR, Greene WA, Choi JH, Wang HC (2015) Deriving retinal pigment epithelium (RPE) from induced pluripotent stem (iPS) cells by different sizes of embryoid bodies. J Vis Exp. https://doi.org/10.3791/52262
36. Kokkinaki M, Sahibzada N, Golestaneh N (2011) Human induced pluripotent stem-derived retinal pigment epithelium (RPE) cells exhibit ion transport, membrane potential, polarized vascular endothelial growth factor secretion, and gene expression pattern similar to native RPE. Stem Cells 29:825–835
37. Muniz A, Greene WA, Plamper ML, Choi JH, Johnson AJ, Tsin AT, Wang HC (2014) Retinoid uptake, processing, and secretion in human iPS-RPE support the visual cycle. Invest Ophthalmol Vis Sci 55:198–209
38. Wang HC, Greene WA, Kaini RR, Shen-Gunther J, Chen HI, Cai H, Wang Y (2014) Profiling the microRNA expression in human iPS and iPS-derived retinal pigment epithelium. Cancer Inform 13:25–35
39. Greene WA, Muniz A, Plamper ML, Kaini RR, Wang HC (2014) MicroRNA expression profiles of human iPS cells, retinal pigment epithelium derived from iPS, and fetal retinal pigment epithelium. J Vis Exp. https://doi.org/10.3791/51589:e51589
40. Li Y, Tsai YT, Hsu CW, Erol D, Yang J, Wu WH, Davis RJ, Egli D, Tsang SH (2012) Long-term safety and efficacy of human-induced pluripotent stem cell (iPS) grafts in a preclinical model of retinitis pigmentosa. Mol Med 18:1312–1319
41. Schwartz SD, Regillo CD, Lam BL, Eliott D, Rosenfeld PJ, Gregori NZ, Hubschman JP, Davis JL, Heilwell G, Spirn M, Maguire J, Gay R, Bateman J, Ostrick RM, Morris D, Vincent M, Anglade E, Del Priore LV, Lanza R (2015) Human embryonic stem cell-derived retinal pigment epithelium in patients with age-related macular degeneration and Stargardt's macular dystrophy: follow-up of two open-label phase 1/2 studies. Lancet 385:509–516
42. Song WK, Park KM, Kim HJ, Lee JH, Choi J, Chong SY, Shim SH, Del Priore LV, Lanza R (2015) Treatment of macular degeneration using embryonic stem cell-derived retinal pigment epithelium: preliminary results in Asian patients. Stem Cell Rep 4:860–872
43. Takahashi K, Tanabe K, Ohnuki M, Narita M, Ichisaka T, Tomoda K, Yamanaka S (2007) Induction of pluripotent stem cells from adult human fibroblasts by defined factors. Cell 131:861–872
44. Wang NK, Tosi J, Kasanuki JM, Chou CL, Kong J, Parmalee N, Wert KJ, Allikmets R, Lai CC, Chien CL, Nagasaki T, Lin CS, Tsang SH (2010) Transplantation of reprogrammed embryonic stem cells improves visual function in a mouse model for retinitis pigmentosa. Transplantation 89:911–919
45. Nguyen HV, Li Y, Tsang SH (2015) Patient-specific iPSC-derived RPE for modeling of retinal diseases. J Clin Med 4:567–578

46. Singh R, Shen W, Kuai D, Martin JM, Guo X, Smith MA, Perez ET, Phillips MJ, Simonett JM, Wallace KA, Verhoeven AD, Capowski EE, Zhang X, Yin Y, Halbach PJ, Fishman GA, Wright LS, Pattnaik BR, Gamm DM (2013) iPS cell modeling of best disease: insights into the pathophysiology of an inherited macular degeneration. Hum Mol Genet 22:593–607

47. Yang J, Li Y, Chan L, Tsai YT, Wu WH, Nguyen HV, Hsu CW, Li X, Brown LM, Egli D, Sparrow JR, Tsang SH (2014) Validation of genome-wide association study (GWAS)-identified disease risk alleles with patient-specific stem cell lines. Hum Mol Genet 23:3445–3455

48. Cereso N, Pequignot MO, Robert L, Becker F, De Luca V, Nabholz N, Rigau V, De Vos J, Hamel CP, Kalatzis V (2014) Proof of concept for AAV2/5-mediated gene therapy in iPSC-derived retinal pigment epithelium of a choroideremia patient. Mol Ther Methods Clin Dev 1:14011

49. Polinati PP, Ilmarinen T, Trokovic R, Hyotylainen T, Otonkoski T, Suomalainen A, Skottman H, Tyni T (2015) Patient-specific induced pluripotent stem cell-derived RPE cells: understanding the pathogenesis of retinopathy in long-chain 3-hydroxyacyl-CoA dehydrogenase deficiency. Invest Ophthalmol Vis Sci 56:3371–3382

50. Song MJ, Bharti K (2016) Looking into the future: using induced pluripotent stem cells to build two and three dimensional ocular tissue for cell therapy and disease modeling. Brain Res 1638:2–14

51. Rehan S, Javaid Z, Al-Bermani A (2015) Unilateral subretinal fibrosis and uveitis syndrome. Scott Med J 60:e4–e6

52. Lobler M, Buss D, Kastner C, Mostertz J, Homuth G, Ernst M, Guthoff R, Wree A, Stahnke T, Fuellen G, Voelker U, Schmitz KP (2013) Ocular fibroblast types differ in their mRNA profiles—implications for fibrosis prevention after aqueous shunt implantation. Mol Vis 19:1321–1331

53. Chen KJ, Sun MH, Lai CC (2012) Massive submacular fibrosis after ocular blunt injury. Arch Ophthalmol 130:1126

54. Bianchi E, Ripandelli G, Feher J, Plateroti AM, Plateroti R, Kovacs I, Plateroti P, Taurone S, Artico M (2015) Occlusion of retinal capillaries caused by glial cell proliferation in chronic ocular inflammation. Folia Morphol (Warsz) 74:33–41

55. Herschler J (1977) Trabecular damage due to blunt anterior segment injury and its relationship to traumatic glaucoma. Trans Sect Ophthalmol Am Acad Ophthalmol Otolaryngol 83:239–248

56. Aylward GW, Lawson J, McCarry B, Lee JP, Fells P (1992) The surgical treatment of traumatic Brown syndrome. J Pediatr Ophthalmol Strabismus 29:276–283

57. Garcia GH, Goldberg RA, Shorr N (1998) The transcaruncular approach in repair of orbital fractures: a retrospective study. J Craniomaxillofac Trauma 4:7–12

58. Dubois L, Steenen SA, Gooris PJ, Bos RR, Becking AG (2016) Controversies in orbital reconstruction-III. Biomaterials for orbital reconstruction: a review with clinical recommendations. Int J Oral Maxillofac Surg 45:41–50

59. Dubois L, Steenen SA, Gooris PJ, Mourits MP, Becking AG (2015) Controversies in orbital reconstruction—II. Timing of post-traumatic orbital reconstruction: a systematic review. Int J Oral Maxillofac Surg 44:433–440

60. Dubois L, Steenen SA, Gooris PJ, Mourits MP, Becking AG (2015) Controversies in orbital reconstruction—I. Defect-driven orbital reconstruction: a systematic review. Int J Oral Maxillofac Surg 44:308–315

61. Mendes S, Campos A, Beselga D, Campos J, Neves A (2014) Traumatic maculopathy 6 months after injury: a clinical case report. Case Rep Ophthalmol 5:78–82

62. Campos J, Campos A, Beselga D, Mendes S, Neves A, Sousa JP (2013) Punctate inner choroidopathy: a clinical case report. Case Rep Ophthalmol 4:155–159

63. Pastor JC (1998) Proliferative vitreoretinopathy: an overview. Surv Ophthalmol 43:3–18

64. Kantelip B, Bacin F (1985) Intraocular fibrosis after perforating injury of the posterior segment. Experimental study. J Fr Ophtalmol 8:245–253

65. Cockerham GC, Rice TA, Hewes EH, Cockerham KP, Lemke S, Wang G, Lin RC, Glynn-Milley C, Zumhagen L (2011) Closed-eye ocular injuries in the Iraq and Afghanistan wars. N Engl J Med 364:2172–2173

66. Moysidis SN, Thanos A, Vavvas DG (2012) Mechanisms of inflammation in proliferative vitreoretinopathy: from bench to bedside. Mediat Inflamm 2012:815937

67. Weichel ED, Colyer MH (2008) Combat ocular trauma and systemic injury. Curr Opin Ophthalmol 19:519–525

68. Pastor JC, de la Rua ER, Martin F (2002) Proliferative vitreoretinopathy: risk factors and pathobiology. Prog Retin Eye Res 21:127–144

69. Snead DR, James S, Snead MP (2008) Pathological changes in the vitreoretinal junction 1: epiretinal membrane formation. Eye (Lond) 22:1310–1317

70. Charteris DG, Sethi CS, Lewis GP, Fisher SK (2002) Proliferative vitreoretinopathy-developments in adjunctive treatment and retinal pathology. Eye (Lond) 16:369–374

71. Erakgun T, Egrilmez S (2009) Surgical outcomes of transconjunctival sutureless 23-gauge vitrectomy with silicone oil injection. Indian J Ophthalmol 57:105–109

72. Shah CP, Ho AC, Regillo CD, Fineman MS, Vander JF, Brown GC (2008) Short-term outcomes of 25-gauge vitrectomy with silicone oil for repair of complicated retinal detachment. Retina 28:723–728

73. Riemann CD, Miller DM, Foster RE, Petersen MR (2007) Outcomes of transconjunctival sutureless 25-gauge vitrectomy with silicone oil infusion. Retina 27:296–303

74. Schaal S, Sherman MP, Barr CC, Kaplan HJ (2011) Primary retinal detachment repair: comparison of 1-year outcomes of four surgical techniques. Retina 31:1500–1504

75. Yao Y, Jiang L, Wang ZJ, Zhang MN (2006) Scleral buckling procedures for longstanding or chronic rhegmatogenous retinal detachment with subretinal proliferation. Ophthalmology 113:821–825

76. Storey P, Alshareef R, Khuthaila M, London N, Leiby B, DeCroos C, Kaiser R, Wills PVRSG (2014) Pars plana vitrectomy and scleral buckle versus pars plana vitrectomy alone for patients with rhegmatogenous retinal detachment at high risk for proliferative vitreoretinopathy. Retina 34:1945–1951

77. Quiram PA, Gonzales CR, Hu W, Gupta A, Yoshizumi MO, Kreiger AE, Schwartz SD (2006) Outcomes of vitrectomy with inferior retinectomy in patients with recurrent rhegmatogenous retinal detachments and proliferative vitreoretinopathy. Ophthalmology 113:2041–2047

78. Tsui I, Schubert HD (2009) Retinotomy and silicone oil for detachments complicated by anterior inferior proliferative vitreoretinopathy. Br J Ophthalmol 93:1228–1233

79. Tan HS, Mura M, Oberstein SY, de Smet MD (2010) Primary retinectomy in proliferative vitreoretinopathy. Am J Ophthalmol 149:447–452

80. Joussen AM, Rizzo S, Kirchhof B, Schrage N, Li X, Lente C, Hilgers RD, Group HSOS (2011) Heavy silicone oil versus standard silicone oil in as vitreous tamponade in inferior PVR (HSO Study): interim analysis. Acta Ophthalmol 89:e483–e489

81. Boscia F, Furino C, Recchimurzo N, Besozzi G, Sborgia G, Sborgia C (2008) Oxane HD vs silicone oil and scleral buckle in retinal detachment with proliferative vitreoretinopathy and inferior retinal breaks. Graefes Arch Clin Exp Ophthalmol 246:943–948

82. Kralinger MT, Stolba U, Velikay M, Egger S, Binder S, Wedrich A, Haas A, Parel JM, Kieselbach GF (2010) Safety and feasibility of a novel intravitreal tamponade using a silicone oil/acetyl-salicylic acid suspension for proliferative vitreoretinopathy: first results of the Austrian Clinical Multicenter Study. Graefes Arch Clin Exp Ophthalmol 248:1193–1198

83. Ahmadieh H, Feghhi M, Tabatabaei H, Shoeibi N, Ramezani A, Mohebbi MR (2008) Triamcinolone acetonide in silicone-filled eyes as adjunctive treatment for proliferative vitreoretinopathy: a randomized clinical trial. Ophthalmology 115:1938–1943

84. Yamakiri K, Sakamoto T, Noda Y, Nakahara M, Ogino N, Kubota T, Yokoyama M, Furukawa M, Ishibashi T (2008) One-year results of a multicenter controlled clinical trial of triamcinolone in pars plana vitrectomy. Graefes Arch Clin Exp Ophthalmol 246:959–966

85. Chen W, Chen H, Hou P, Fok A, Hu Y, Lam DS (2011) Midterm results of low-dose intravitreal triamcinolone as adjunctive treatment for proliferative vitreoretinopathy. Retina 31:1137–1142

86. Dehghan MH, Ahmadieh H, Soheilian M, Azarmina M, Moradian S, Ramezani AR, Tavallal A, Naghibozakerin J (2010) Effect of oral prednisolone on visual outcomes and complications after scleral buckling. Eur J Ophthalmol 20:419–423

87. Reibaldi M, Russo A, Longo A, Bonfiglio V, Uva MG, Gagliano C, Toro MD, Avitabile T (2013) Rhegmatogenous retinal detachment with a high risk of proliferative vitreoretinopathy treated with episcleral surgery and an intravitreal dexamethasone 0.7-mg implant. Case Rep Ophthalmol 4:79–83

88. Banerjee PJ, Bunce C, Charteris DG (2013) Ozurdex (a slow-release dexamethasone implant) in proliferative vitreoretinopathy: study protocol for a randomised controlled trial. Trials 14:358

89. Kumar A, Nainiwal S, Choudhary I, Tewari HK, Verma LK (2002) Role of daunorubicin in inhibiting proliferative vitreoretinopathy after retinal detachment surgery. Clin Exp Ophthalmol 30:348–351

90. Wiedemann P, Hilgers RD, Bauer P, Heimann K (1998) Adjunctive daunorubicin in the treatment of proliferative vitreoretinopathy: results of a multicenter clinical trial. Daunomycin Study Group. Am J Ophthalmol 126:550–559

91. Asaria RH, Kon CH, Bunce C, Charteris DG, Wong D, Khaw PT, Aylward GW (2001) Adjuvant 5-fluorouracil and heparin prevents proliferative vitreoretinopathy: results from a randomized, double-blind, controlled clinical trial. Ophthalmology 108:1179–1183

92. Charteris DG, Aylward GW, Wong D, Groenewald C, Asaria RH, Bunce C, Group PVRS (2004) A randomized controlled trial of combined 5-fluorouracil and low-molecular-weight heparin in management of established proliferative vitreoretinopathy. Ophthalmology 111:2240–2245

93. Wickham L, Bunce C, Wong D, McGurn D, Charteris DG (2007) Randomized controlled trial of combined 5-fluorouracil and low-molecular-weight heparin in the management of unselected rhegmatogenous retinal detachments undergoing primary vitrectomy. Ophthalmology 114:698–704

94. Fekrat S, de Juan E Jr, Campochiaro PA (1995) The effect of oral 13-cis-retinoic acid on retinal redetachment after surgical repair in eyes with proliferative vitreoretinopathy. Ophthalmology 102:412–418

95. Chang YC, Hu DN, Wu WC (2008) Effect of oral 13-cis-retinoic acid treatment on postoperative clinical outcome of eyes with proliferative vitreoretinopathy. Am J Ophthalmol 146:440–446

96. Hsu J, Khan MA, Shieh WS, Chiang A, Maguire JI, Park CH, Garg SJ, Ho AC, Kaiser RS (2016) Effect of serial intrasilicone oil bevacizumab injections in eyes with recurrent proliferative vitreoretinopathy retinal detachment. Am J Ophthalmol 161:65–70.e62

97. Mandava N, Blackburn P, Paul DB, Wilson MW, Read SB, Alspaugh E, Tritz R, Barber JR, Robbins JM, Kruse CA (2002) Ribozyme to proliferating cell nuclear antigen to treat proliferative vitreoretinopathy. Invest Ophthalmol Vis Sci 43:3338–3348

98. Schiff WM, Hwang JC, Ober MD, Olson JL, Dhrami-Gavazi E, Barile GR, Chang S, Mandava N (2007) Safety and efficacy assessment of chimeric ribozyme to proliferating cell nuclear antigen to prevent recurrence of proliferative vitreoretinopathy. Arch Ophthalmol 125:1161–1167

99. Khan MA, Brady CJ, Kaiser RS (2015) Clinical management of proliferative vitreoretinopathy: an update. Retina 35:165–175

100. Charteris DG (1995) Proliferative vitreoretinopathy: pathobiology, surgical management, and adjunctive treatment. Br J Ophthalmol 79:953–960

101. Garweg JG, Tappeiner C, Halberstadt M (2013) Pathophysiology of proliferative vitreoretinopathy in retinal detachment. Surv Ophthalmol 58:321–329

102. Pennock S, Haddock LJ, Eliott D, Mukai S, Kazlauskas A (2014) Is neutralizing vitreal growth factors a viable strategy to prevent proliferative vitreoretinopathy? Prog Retin Eye Res 40:16–34

103. Nagasaki H, Shinagawa K, Mochizuki M (1998) Risk factors for proliferative vitreoretinopathy. Prog Retin Eye Res 17:77–98

104. Pastor JC, Rojas J, Pastor-Idoate S, Di Lauro S, Gonzalez-Buendia L, Delgado-Tirado S (2015) Proliferative vitreoretinopathy: a new concept of disease pathogenesis and practical consequences. Prog Retin Eye Res 51:125. https://doi.org/10.1016/j.preteyeres.2015.07.005

105. Laqua H, Machemer R (1975) Glial cell proliferation in retinal detachment (massive periretinal proliferation). Am J Ophthalmol 80:602–618

106. Machemer R, Laqua H (1975) Pigment epithelium proliferation in retinal detachment (massive periretinal proliferation). Am J Ophthalmol 80:1–23

107. Laqua H (1975) Massive periretinal proliferation (MPP) IV. Pre- and subretinal proliferation of glial tissue in experimental retinal detachment. Mod Probl Ophthalmol 15:235–245

108. Baudouin C, Hofman P, Brignole F, Bayle J, Loubiere R, Gastaud P (1991) Immunocytology of cellular components in vitreous and subretinal fluid from patients with proliferative vitreoretinopathy. Ophthalmologica 203:38–46

109. Wiedemann P, Weller M (1988) The pathophysiology of proliferative vitreoretinopathy. Acta Ophthalmol Suppl 189:3–15

110. Charteris DG, Hiscott P, Grierson I, Lightman SL (1992) Proliferative vitreoretinopathy. Lymphocytes in epiretinal membranes. Ophthalmology 99:1364–1367

111. Charteris DG, Hiscott P, Robey HL, Gregor ZJ, Lightman SL, Grierson I (1993) Inflammatory cells in proliferative vitreoretinopathy subretinal membranes. Ophthalmology 100:43–46

112. Morescalchi F, Duse S, Gambicorti E, Romano MR, Costagliola C, Semeraro F (2013) Proliferative vitreoretinopathy after eye injuries: an overexpression of growth factors and cytokines leading to a retinal keloid. Mediat Inflamm 2013:269787

113. Casaroli-Marano RP, Pagan R, Vilaro S (1999) Epithelial-mesenchymal transition in proliferative vitreoretinopathy: intermediate filament protein expression in retinal pigment epithelial cells. Invest Ophthalmol Vis Sci 40:2062–2072

114. Wu WC, Kao YH, Hu DN (2000) Relationship between outcome of proliferative vitreoretinopathy and results of tissue culture of excised preretinal membranes. Kaohsiung J Med Sci 16:614–619

115. Anderson DH, Stern WH, Fisher SK, Erickson PA, Borgula GA (1983) Retinal detachment in the cat: the pigment epithelial-photoreceptor interface. Invest Ophthalmol Vis Sci 24:906–926

116. Lee SC, Kwon OW, Seong GJ, Kim SH, Ahn JE, Kay ED (2001) Epitheliomesenchymal transdifferentiation of cultured RPE cells. Ophthalmic Res 33:80–86

117. Stocks SZ, Taylor SM, Shiels IA (2001) Transforming growth factor-beta1 induces alpha-smooth muscle actin expression and fibronectin synthesis in cultured human retinal pigment epithelial cells. Clin Exp Ophthalmol 29:33–37

118. Glaser BM, Cardin A, Biscoe B (1987) Proliferative vitreoretinopathy. The mechanism of development of vitreoretinal traction. Ophthalmology 94:327–332

119. Hiscott P, Sheridan C, Magee RM, Grierson I (1999) Matrix and the retinal pigment epithelium in proliferative retinal disease. Prog Retin Eye Res 18:167–190

120. Agrawal RN, He S, Spee C, Cui JZ, Ryan SJ, Hinton DR (2007) In vivo models of proliferative vitreoretinopathy. Nat Protoc 2:67–77

121. Chiba C (2014) The retinal pigment epithelium: an important player of retinal disorders and regeneration. Exp Eye Res 123:107–114

122. Guerin CJ, Hu L, Scicli G, Scicli AG (2001) Transforming growth factor beta in experimentally detached retina and periretinal membranes. Exp Eye Res 73:753–764

123. Hinton DR, He S, Jin ML, Barron E, Ryan SJ (2002) Novel growth factors involved in the pathogenesis of proliferative vitreoretinopathy. Eye (Lond) 16:422–428

124. Elner SG, Elner VM, Jaffe GJ, Stuart A, Kunkel SL, Strieter RM (1995) Cytokines in proliferative diabetic retinopathy and proliferative vitreoretinopathy. Curr Eye Res 14:1045–1053

125. Desmouliere A, Geinoz A, Gabbiani F, Gabbiani G (1993) Transforming growth factor-beta 1 induces alpha-smooth muscle actin expression in granulation tissue myofibroblasts and in quiescent and growing cultured fibroblasts. J Cell Biol 122:103–111

126. Desmouliere A, Gabbiani G (1995) Myofibroblast differentiation during fibrosis. Exp Nephrol 3:134–139

127. Desmouliere A (1995) Factors influencing myofibroblast differentiation during wound healing and fibrosis. Cell Biol Int 19:471–476
128. Tomasek JJ, Gabbiani G, Hinz B, Chaponnier C, Brown RA (2002) Myofibroblasts and mechano-regulation of connective tissue remodelling. Nat Rev Mol Cell Biol 3:349–363
129. Gabbiani G (2003) The myofibroblast in wound healing and fibrocontractive diseases. J Pathol 200:500–503
130. Baudouin C, Fredj-Reygrobellet D, Brignole F, Negre F, Lapalus P, Gastaud P (1993) Growth factors in vitreous and subretinal fluid cells from patients with proliferative vitreoretinopathy. Ophthalmic Res 25:52–59
131. Cassidy L, Barry P, Shaw C, Duffy J, Kennedy S (1998) Platelet derived growth factor and fibroblast growth factor basic levels in the vitreous of patients with vitreoretinal disorders. Br J Ophthalmol 82:181–185
132. Lei H, Hovland P, Velez G, Haran A, Gilbertson D, Hirose T, Kazlauskas A (2007) A potential role for PDGF-C in experimental and clinical proliferative vitreoretinopathy. Invest Ophthalmol Vis Sci 48:2335–2342
133. Pennock S, Rheaume MA, Mukai S, Kazlauskas A (2011) A novel strategy to develop therapeutic approaches to prevent proliferative vitreoretinopathy. Am J Pathol 179:2931–2940
134. Cui J, Lei H, Samad A, Basavanthappa S, Maberley D, Matsubara J, Kazlauskas A (2009) PDGF receptors are activated in human epiretinal membranes. Exp Eye Res 88:438–444
135. Lee H, O'Meara SJ, O'Brien C, Kane R (2007) The role of gremlin, a BMP antagonist, and epithelial-to-mesenchymal transition in proliferative vitreoretinopathy. Invest Ophthalmol Vis Sci 48:4291–4299
136. Hill TP, Spater D, Taketo MM, Birchmeier W, Hartmann C (2005) Canonical Wnt/beta-catenin signaling prevents osteoblasts from differentiating into chondrocytes. Dev Cell 8:727–738
137. Umazume K, Tsukahara R, Liu L, Fernandez de Castro JP, McDonald K, Kaplan HJ, Tamiya S (2014) Role of retinal pigment epithelial cell beta-catenin signaling in experimental proliferative vitreoretinopathy. Am J Pathol 184:1419–1428
138. Chen Z, Shao Y, Li X (2015) The roles of signaling pathways in epithelial-to-mesenchymal transition of PVR. Mol Vis 21:706–710
139. Yang S, Li H, Li M, Wang F (2015) Mechanisms of epithelial-mesenchymal transition in proliferative vitreoretinopathy. Discov Med 20:207–217
140. Chen X, Xiao W, Wang W, Luo L, Ye S, Liu Y (2014) The complex interplay between ERK1/2, TGFbeta/Smad, and Jagged/Notch signaling pathways in the regulation of epithelial-mesenchymal transition in retinal pigment epithelium cells. PLoS One 9:e96365
141. Parrales A, Lopez E, Lee-Rivera I, Lopez-Colome AM (2013) ERK1/2-dependent activation of mTOR/mTORC1/p70S6K regulates thrombin-induced RPE cell proliferation. Cell Signal 25:829–838
142. Yokoyama K, Kimoto K, Itoh Y, Nakatsuka K, Matsuo N, Yoshioka H, Kubota T (2012) The PI3K/Akt pathway mediates the expression of type I collagen induced by TGF-beta2 in human retinal pigment epithelial cells. Graefes Arch Clin Exp Ophthalmol 250:15–23
143. Cheng HC, Ho TC, Chen SL, Lai HY, Hong KF, Tsao YP (2008) Troglitazone suppresses transforming growth factor beta-mediated fibrogenesis in retinal pigment epithelial cells. Mol Vis 14:95–104
144. Li H, Wang H, Wang F, Gu Q, Xu X (2011) Snail involves in the transforming growth factor beta1-mediated epithelial-mesenchymal transition of retinal pigment epithelial cells. PLoS One 6:e23322
145. Gamulescu MA, Chen Y, He S, Spee C, Jin M, Ryan SJ, Hinton DR (2006) Transforming growth factor beta2-induced myofibroblastic differentiation of human retinal pigment epithelial cells: regulation by extracellular matrix proteins and hepatocyte growth factor. Exp Eye Res 83:212–222
146. Lei H, Rheaume MA, Kazlauskas A (2010) Recent developments in our understanding of how platelet-derived growth factor (PDGF) and its receptors contribute to proliferative vitreoretinopathy. Exp Eye Res 90:376–381

147. Li M, Li H, Liu X, Xu D, Wang F (2014) MicroRNA-29b regulates TGF-beta1-mediated epithelial-mesenchymal transition of retinal pigment epithelial cells by targeting AKT2. Exp Cell Res 345:115. https://doi.org/10.1016/j.yexcr.2014.09.026

148. Palma-Nicolas JP, Lopez-Colome AM (2013) Thrombin induces slug-mediated E-cadherin transcriptional repression and the parallel up-regulation of N-cadherin by a transcription-independent mechanism in RPE cells. J Cell Physiol 228:581–589

149. Bastiaans J, van Meurs JC, van Holten-Neelen C, Nagtzaam NM, van Hagen PM, Chambers RC, Hooijkaas H, Dik WA (2013) Thrombin induces epithelial-mesenchymal transition and collagen production by retinal pigment epithelial cells via autocrine PDGF-receptor signaling. Invest Ophthalmol Vis Sci 54:8306–8314

150. Bastiaans J, van Meurs JC, van Holten-Neelen C, Nijenhuis MS, Kolijn-Couwenberg MJ, van Hagen PM, Kuijpers RW, Hooijkaas H, Dik WA (2013) Factor Xa and thrombin stimulate proinflammatory and profibrotic mediator production by retinal pigment epithelial cells: a role in vitreoretinal disorders? Graefes Arch Clin Exp Ophthalmol 251:1723–1733

151. Lei H, Kazlauskas A (2014) A reactive oxygen species-mediated, self-perpetuating loop persistently activates platelet-derived growth factor receptor alpha. Mol Cell Biol 34:110–122

152. Chen HC, Zhu YT, Chen SY, Tseng SC (2012) Wnt signaling induces epithelial-mesenchymal transition with proliferation in ARPE-19 cells upon loss of contact inhibition. Lab Investig 92:676–687

153. Kita T, Hata Y, Miura M, Kawahara S, Nakao S, Ishibashi T (2007) Functional characteristics of connective tissue growth factor on vitreoretinal cells. Diabetes 56:1421–1428

154. Chen YJ, Tsai RK, Wu WC, He MS, Kao YH, Wu WS (2012) Enhanced PKCdelta and ERK signaling mediate cell migration of retinal pigment epithelial cells synergistically induced by HGF and EGF. PLoS One 7:e44937

155. Pennock S, Haddock LJ, Mukai S, Kazlauskas A (2014) Vascular endothelial growth factor acts primarily via platelet-derived growth factor receptor alpha to promote proliferative vitreoretinopathy. Am J Pathol 184:3052–3068

156. Carrington L, McLeod D, Boulton M (2000) IL-10 and antibodies to TGF-beta2 and PDGF inhibit RPE-mediated retinal contraction. Invest Ophthalmol Vis Sci 41:1210–1216

157. Connor TB Jr, Roberts AB, Sporn MB, Danielpour D, Dart LL, Michels RG, de Bustros S, Enger C, Kato H, Lansing M et al (1989) Correlation of fibrosis and transforming growth factor-beta type 2 levels in the eye. J Clin Invest 83:1661–1666

158. Kita T, Hata Y, Arita R, Kawahara S, Miura M, Nakao S, Mochizuki Y, Enaida H, Goto Y, Shimokawa H, Hafezi-Moghadam A, Ishibashi T (2008) Role of TGF-beta in proliferative vitreoretinal diseases and ROCK as a therapeutic target. Proc Natl Acad Sci U S A 105:17504–17509

159. Chen X, Xiao W, Liu X, Zeng M, Luo L, Wu M, Ye S, Liu Y (2014) Blockade of Jagged/Notch pathway abrogates transforming growth factor beta2-induced epithelial-mesenchymal transition in human retinal pigment epithelium cells. Curr Mol Med 14:523–534

160. Sonoda S, Nagineni CN, Kitamura M, Spee C, Kannan R, Hinton DR (2014) Ceramide inhibits connective tissue growth factor expression by human retinal pigment epithelial cells. Cytokine 68:137–140

161. Liang CM, Tai MC, Chang YH, Chen YH, Chen CL, Lu DW, Chen JT (2011) Glucosamine inhibits epithelial-to-mesenchymal transition and migration of retinal pigment epithelium cells in culture and morphologic changes in a mouse model of proliferative vitreoretinopathy. Acta Ophthalmol 89:e505–e514

162. Choi K, Lee K, Ryu SW, Im M, Kook KH, Choi C (2012) Pirfenidone inhibits transforming growth factor-beta1-induced fibrogenesis by blocking nuclear translocation of Smads in human retinal pigment epithelial cell line ARPE-19. Mol Vis 18:1010–1020

163. Oshima Y, Sakamoto T, Hisatomi T, Tsutsumi C, Ueno H, Ishibashi T (2002) Gene transfer of soluble TGF-beta type II receptor inhibits experimental proliferative vitreoretinopathy. Gene Ther 9:1214–1220

164. Lei H, Velez G, Hovland P, Hirose T, Gilbertson D, Kazlauskas A (2009) Growth factors outside the PDGF family drive experimental PVR. Invest Ophthalmol Vis Sci 50:3394–3403

165. Xiao W, Chen X, Liu X, Luo L, Ye S, Liu Y (2014) Trichostatin A, a histone deacetylase inhibitor, suppresses proliferation and epithelial-mesenchymal transition in retinal pigment epithelium cells. J Cell Mol Med 18:646–655
166. Lei H, Velez G, Cui J, Samad A, Maberley D, Matsubara J, Kazlauskas A (2010) N-acetylcysteine suppresses retinal detachment in an experimental model of proliferative vitreoretinopathy. Am J Pathol 177:132–140
167. Kusaka K, Kothary PC, Del Monte MA (1998) Modulation of basic fibroblast growth factor effect by retinoic acid in cultured retinal pigment epithelium. Curr Eye Res 17:524–530
168. Umazume K, Liu L, Scott PA, de Castro JP, McDonald K, Kaplan HJ, Tamiya S (2013) Inhibition of PVR with a tyrosine kinase inhibitor, dasatinib, in the swine. Invest Ophthalmol Vis Sci 54:1150–1159

Chapter 3
Developing Cell-Based Therapies for RPE-Associated Degenerative Eye Diseases

Karim Ben M'Barek, Walter Habeler, Florian Regent,
and Christelle Monville

Abstract In developed countries, blindness and visual impairment are caused mainly by diseases affecting the retina. These retinal degenerative diseases, including age-related macular dystrophy (AMD) and inherited retinal diseases such as retinitis pigmentosa (RP), are the predominant causes of human blindness worldwide and are responsible for more than 1.5 million cases in France and more than 30 million cases worldwide. Global prevalence and disease burden projections for next 20 years are alarming (Wong et al., Lancet Glob Health 2(2):e106–e116, 2014) and strongly argue toward designing innovative eye-care strategies. At present, despite the scientific advances achieved in the last years, there is no cure for such diseases, making retinal degenerative diseases an unmet medical need.

The majority of the inherited retinal disease (IRD) genes codes for proteins acting directly in photoreceptors. Yet, a few of them are expressed in the retinal pigment epithelium (RPE), the supporting tissue necessary for proper functioning of the photoreceptors. Among retinal degenerative diseases, impairment of some RPE genes engenders a spectrum of conditions ranging from stationary visual defects to very severe forms of retinal dystrophies in which the RPE dysfunction leads to photoreceptors cell death and consecutive irreversible vision loss. The accessibility of the eye and the immune privilege of the retina, together with the availability of noninvasive imaging technologies, make such inherited retinal dystrophies a

K. Ben M'Barek · W. Habeler
INSERM U861, I-Stem, AFM, Institute for Stem Cell Therapy and Exploration
of Monogenic Diseases, Corbeil-Essonnes, France

UEVE UMR861, Corbeil-Essonnes, France

CECS, Association Française contre les Myopathies, Corbeil-Essonnes, France

F. Regent · C. Monville (✉)
INSERM U861, I-Stem, AFM, Institute for Stem Cell Therapy and Exploration
of Monogenic Diseases, Corbeil-Essonnes, France

UEVE UMR861, Corbeil-Essonnes, France
e-mail: CMONVILLE@istem.fr

© Springer Nature Switzerland AG 2019
K. Bharti (ed.), *Pluripotent Stem Cells in Eye Disease Therapy*,
Advances in Experimental Medicine and Biology 1186,
https://doi.org/10.1007/978-3-030-28471-8_3

particularly attractive disease model for innovative cell therapy approaches to replace, regenerate, and/or repair the injured RPE tissue. Proof-of-concept studies in animal models have demonstrated the safety and efficacy of the engraftment of therapeutic cells either to support RPE cell functions or to provide a trophic support to photoreceptors. These different approaches are now in the pipeline of drug development with objective to provide first cell-based treatments by 2020.

This chapter will focus on the different cell-based strategies developed in the past and current approaches to prevent photoreceptor death in RPE-associated degenerative eye diseases.

Keywords RPE · Human pluripotent stem cells · Cell therapy · Clinical trials · Inherited retinal diseases

3.1   Retinopathies Induced by a Primary RPE Defect

The retina is a 500 µm-thick light-sensitive tissue lining the inner surface of the eye. It is formed by multiple layers of interconnected neurons and is in charge of the first steps of visual processing (Fig. 3.1). The photoreceptors (rods and cones) are, together with the ganglion cells, the photosensitive cells of the retina. Rods and cones initiate the conversion of light energy into electrical signals through a process called phototransduction. Retinal interneurons (amacrine, bipolar, and horizontal cells) further codify the electrical signals into optic nerve impulses (through ganglion cells), which are subsequently interpreted by the brain as visual images [2]. Underlying the photoreceptors layer, we could find the RPE, constituted by a polarized monolayer of epithelial cells, which are real babysitters for photoreceptors. The RPE cells display a hexagonal shape and form a pigmented epithelium that lies between the choroid and the neural retina (Fig. 3.1). It is flanked by the Bruch's membrane on its basal surface and by the outer segments of the photoreceptors on its apical portion. The cells in this layer are connected by tight junctions and constitute the outer components of the blood–retinal barrier [3, 4]. In the subretinal space (SRS) filled with the interphotoreceptor matrix, microvilli from RPE cells appose to the photoreceptor outer segments, forming the physical components that can contribute to the maintenance of retinal adhesion [5].

RPE cells perform a number of functions that are critical for the survival and proper functioning of the photoreceptors [6]. RPE cells, among other functions, are involved in the recycling of the vitamin A, buffer ion composition in the SRS, secrete growth factors, regulate T-cell activation in the eye, and phagocytize shed outer segment of photoreceptors [7, 8]. All these functions are essential, and a failure of any of them can lead to photoreceptors death, degeneration of the retina,

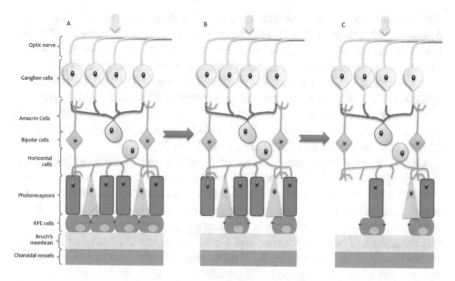

Fig. 3.1 Structure of normal retina (**a**) and after RPE and photoreceptor degeneration (**b, c**)

loss of visual function, and ultimately blindness. Several disease conditions specifically alter RPE cells with major consequences in vision.

3.1.1 Retinitis Pigmentosa

RP is a highly heterogeneous group of retinal diseases affecting either photoreceptors or RPE or both [4, 9, 10] with incidence of 1/3500–4000 [11]. These diseases are rare monogenic dystrophies. Some mutations specifically affect RPE genes (https://sph.uth.edu/retnet/disease.htm) such as lecithin retinol transferase (*LRAT*), retinal pigment epithelium-specific 65-kDa protein (*RPE65*), the MER tyrosine kinase proto-oncogene (*MERTK*) and bestrophin (*BEST1*) [4]. They are mainly inherited autosomal recessive, except for some dominant mutations on the bestrophin gene [12]. Though RP has a highly variable clinical presentation and progression, the majority of patients initially experience problems in night vision, since the rod photoreceptors are typically damaged first (Fig. 3.1). Then the peripheral vision is progressively lost leading to tunnel vision. In many cases, this can progress to include the central visual field and blindness [11].

Impairment of some RPE genes involved in the visual cycle leads to a spectrum of conditions ranging from stationary visual defects to very severe forms of retinal dystrophies. Additionally, in RPE disorders, pigment deposits in retina are scarce, even at late disease stages. This feature could reflect impaired RPE migration and proliferation, while with primary defects in photoreceptor cells, the RPE could be healthier. *RPE65* was the first RPE gene reported to be involved in human retinal dystrophy. Mutations in *RPE65* are responsible for 5–10% of the cases of Leber congenital amaurosis (LCA) [13] and a few cases (1–2%) of recessive RP [14–17]. The condition is usually severe and congenital. However, milder phenotypes could be encountered and reported as RP or retinal dystrophy [17, 18]. In contrast to *RPE65*, *LRAT* mutations are a very rare cause of severe retinal dystrophy. The phenotypes of *CRALBP* (*RLBP1*) and *RDH5* mutations are quite similar, although with some differences. For both genes, night blindness is present from the first years of life and dark adaptation thresholds are elevated [19, 20].

One of the most important functions of the RPE is to participate actively in the photoreceptor membrane turnover. It has been calculated that each rod regenerates its outer segment (ROS) within 7–12 days [21]. Photoreceptor membrane turnover is composed of two phases, necessarily balanced in order to preserve cellular integrity: an anabolic aspect involving RNA transcription, protein synthesis, transport, and generation of new membranes; and a catabolic aspect consisting of exhausted membrane removal, digestion by RPE cells (phagocytosis), and recycling of some components to the photoreceptors [22]. Different receptors were identified to have a role in photoreceptor membrane phagocytosis by RPE cells: a mannose-6-phosphate receptor [23], CD36 [24, 25], $\alpha V \beta 5$ integrin [26], and a number of glycoproteins expressed on the RPE surface [27]. A real breakthrough was the identification of the c-mer transmembrane tyrosine kinase (MERTK) receptor as responsible for photoreceptor outer segment (POS) internalization [28, 29]. Mutations in the *MERTK* gene have been linked to subgroups of patients with inherited retinal degeneration [30, 31].

Macular dystrophies are inherited retinal dystrophies in which various forms of deposits; pigmentary changes and atrophic lesions are observed in the macula lutea, the cone-rich region of the human central retina. Mutations in several genes, expressed in RPE cells (*BEST1*, *CDH3*, *TIMP3*), are associated with these conditions [14].

3.1.2   Age-Related Macular Degeneration

AMD is the leading cause of visual impairment in western countries in patients over 50 years of age. The prevalence of 0.05% before the age of 50 years increases to about 12% in population older than 80 years old. While the disease is particularly prevalent in developed countries, it is becoming a global health concern due to the worldwide increase in life expectancies [1]. Besides age, family history has become the second largest risk factor, and smoking, diet, obesity, hypertension, and chronic inflammation have been reported as other environmental risk factors. Although the primary site of the pathologic insult in AMD is not completely determined, different observations support the involvement of oxidative damages, inflammatory changes, and the gradual accumulation of deposits within the RPE cells [32].

Early AMD is characterized by the thickening of Bruch's membrane, deposits under the RPE layer, and by pigmentation changes in the macula. More advanced stages of the disease demonstrate either subretinal neovascularization or atrophy of the retina and RPE. Based on the absence or presence of abnormal choroidal blood vessels' growth below the RPE layer, two different forms of AMD can be distinguished; (1) the 'dry' or geographic atrophy form, the most frequent one (around 90% of AMD patients), resulting from RPE dysfunction leading to accumulation of pigment debris and drusen (deposits of proteins and lipids) and slowly progressing photoreceptor cell death, and (2) the 'wet' or choroidal neovascular form arising from a more rapid and destructive process of choroidal neovascularization through the Bruch's membrane and the RPE in the SRS accompanied by exudation and hemorrhage [33]. The resulting damage to the macular region leads to a loss of vision in the central visual field.

With age, the eye undergoes several changes, some of which affect the RPE. The accumulation of lipofuscin, a reduction in melanin, diminished antioxidant capacity, and the progressive accumulation of deposits underlying Bruch's membrane are hallmarks of aging eyes [3, 34]. Acute and chronic progressive dysfunctions of RPE cells and the age-related deterioration of this tissue have been shown to play a relevant role in AMD [5]. The inflammatory responses observed in AMD retinas are similar but more severe than those observed in normal aging [35]. Moreover, different lifestyle factors, environmental conditions, and gene alterations are likely to explain why certain individuals develop AMD with age, while others do not [36].

In both the atrophic and neovascular forms of AMD, the mutualistic and symbiotic relationship between the photoreceptors, RPE, Bruch's membrane, and choriocapillaris is lost, which results in the death and dysfunction of all the components in the complex [37, 38]. Emerging evidences show that both "dry" AMD and "wet" AMD involve a common pathophysiological background and share similar initiating molecular and cellular alterations [39]. However, the differentiation of dry AMD and wet AMD is clinically relevant because of their distinct clinical presentation and therapeutic options.

3.1.3 Global Burden and Treatments

Altogether, diseases caused by RPE malfunction affect at least 30 millions of people worldwide, a figure that will triple with the increase in the aging population in the next 30–40 years [1]. In addition, as the population is growing older and higher expectations of better quality of life including ability of driving and reading are being asked by patients, morbidity resulting from AMD is becoming increasingly significant. A recent systematic review and meta-analysis has shown that 8.7% of the worldwide population has age-related macular degeneration, and the projected number of people with the disease is around 196 million in 2020, increasing to 288 million in 2040 [1].

Despite a crucial need for therapies, there is no treatment approved for RP. For AMD, due to the complex multifactorial nature of the disease, therapeutic options have evolved slowly. Lifestyle modifications, including smoking cessation and high-dose antioxidant supplements, were shown to decrease disease progression rate. Photocoagulation ablation of extrafoveal new vessels was the first treatment proposed by clinicians for wet AMD and is still in used in selected cases [40]. Photodynamic therapy with systemic administration of a photosensitizing agent and more selective ablation of abnormal vessels were also tested, but recovery was not complete and high rate of recurrence was observed [41]. Anti-VEGF (Vascular Epithelial Growth Factor) agents, which can prevent new vessel development, have significantly improved the prognosis of wet AMD. Repeated injections of these agents formulated by different companies, such as Pegaptanib (Macugen; Eyetech/Pfizer) [42], Ranibizumab (Lucentis; Genentech) [43–45], Aflibercept (Eylea; Regeneron) [46], or Bevacizumab (Avastin; Genentech) [47–49] show the capacity to slow the rate of vision loss but have no more than a 30% rate of effectiveness in all AMD cases.

Emerging treatment modalities include RNA interferences and attempt to block pathways down- or upstream of VEGF to prevent new vessel growth. Anti-PDGF (Platelet-Derived Growth Factor) agents (FovistaTM; Ophthotech, USA) that block pericytes' recruitment were tried in combination with anti-VEGF medications [50].

While promising, gene therapy aiming at restoring gene defects in the RPE (like *RPE65*) shows limitations. Indeed, the prerequisite to this type of gene therapy is to know precisely the defective gene, and a second limitation is to develop a gene delivery candidate drug for each one of the defective genes. Moreover, recent data from preclinical and clinical studies show that, despite stabilization of vision during the follow-up period, patients treated by RPE65 gene therapy still undergo photoreceptor degeneration [51].

Although prognosis of wet AMD has been improved during the last decade, many patients may lose their sight because of late diagnosis or inadequate treatment. For dry AMD, no treatments are available. Different treatment strategies for both AMD could be envisioned: (1) preventing RPE dysfunction or death [52, 53], (2) providing support to stressed RPE [10], and finally (3) replacing RPE and eventually photoreceptors [54].

3.2   Therapeutics Based on Fetal or Adult Cells

The treatment of RPE-associated degenerative eye diseases by cell therapy was first proposed by Gouras and collaborators in early 1980s [55, 56]. This group successfully transplanted RPE cells in a monkey model, demonstrating the feasibility of such approach. By the end of the 1980s, it was well established that RPE cells can not only be transplanted in degenerated retina but also can delay photoreceptor degeneration in a rat model of retinal degeneration [57], opening the path to the development of therapeutic cell-based strategies. Laboratories from all over the world developed and tested methods to produce cells that can replace defective or degenerated endogenous RPE cells.

3.2.1 Use of Fetal-Derived Cells to Replace RPE Cells

One of the first sources identified was the fetal RPE cells. However, the use of such cells required ethical agreement from the civil society and is legally regulated in each country. Researchers and physicians agree to the ethical tenets that regulate experimentation requiring human material including informed consent, no incentive for abortion, procurement of human material that, if not used for research, would be discarded, and review of the protocol from ethical committees. In addition, in case of a clinical use of fetal cells, a detailed medical history of the donor and a complete separation of donor and recipient are required. This set of principles was adopted by the World Medical Association in the declaration of Helsinki (now in its seventh revision) [58].

Human fetal RPE cells from eyes of 10–21 weeks gestational age are separated from the anterior segment and the neural retina by dissection [59–61]. Then RPE cells can be either mechanically dissected or enzymatically separated in order to have an enriched cell culture [60, 61]. After 1 week, a mean of 1.4×10^6 RPE cells can be obtained and passaged for up to ten times [61]. Another protocol from the National Eye Institute (NIH, Bethesda, MA) allows obtaining between 160 and 250 millions of fetal RPE at Passage 1 (P1) per eye, increasing thus the yield [60].

To test the potential for cell therapy of fetal cells, the most useful model is the Royal College of Surgeons (RCS) rat. This model is characterized by a mutation in the *MERTK* gene coding for the transmembrane receptor tyrosine kinase MERTK [62]. In RCS rats, RPE are not able to phagocytosis photoreceptor outer segments [63]. This dysfunction leads to an accumulation of debris between RPE and photoreceptors. As the mutation affects primarily RPE cells, the RCS rat is a good model for both AMD and RP in which the mutation affects RPE cell functions [28, 30]. The photoreceptors begin to die from the third postnatal week. The degeneration progresses quickly, and by 2 months of age, the electroretinogram (ERG) is almost flat in response to light. It indicates that the retina is no more sensitive to light due to the photoreceptors' degeneration. In addition, the decline in vision of RCS rats is also measured in the optokinetic paradigm. In that test, the animal is placed in a platform and exposed to moving stripes. When the rat detects the stimulus, a reflex movement of the head is observed. By modulating the size of stripes, a visual acuity is measured corresponding to the limit of the stimulus detection. In RCS rat, the visual acuity decreases drastically by 2 months of age.

In 1996, Little and collaborators have clearly demonstrated in RCS rats that human fetal RPE cells harvested from enucleated eyes of 10–16 weeks gestational age fetuses have a rescue effect [64]. Indeed 24-day-old rats were grafted with these fetal RPE cells and were sacrificed 1 month later. They reported, after histological examination, that photoreceptors were preserved in the areas surrounding the grafting site compared to sham injected site. They confirmed the presence of the cells by analysis of pigmented cells in the nonpigmented rat eye. RCS rats were immunocompromised by a daily injection of cyclosporine. This treatment was sufficient to prevent rejection during the follow-up.

Another strategy consisted of grafting small patches of fetal RPE still organized as a monolayer (1–5 mm^2) [65]. Transplanted fetal RPE was rejected within 1 month into the rabbit eye but survived at 3 months in the monkey retina without immunosuppression. Moreover grafted RPE is able to phagocytosis outer segments as demonstrated by the presence of phagosomes in electron micrographies of rabbit retinas [65]. Berglin and collaborators had grafted human fetal retinas in six monkeys without immunosuppression. Integrity of local photoreceptors and survival of RPE patch transplants were effective for at least 6 months [66]. Interestingly rejections occur more frequently around fovea (60%) than at the periphery (30%). Prevention of rejection can be managed by Cyclosporin A treatment into the vitreous [67]. Lai's report compared intravitreal release of Cyclosporine A from a capsule sutured in the vitreous or through weekly injections in rabbits transplanted with fetal RPE cell suspensions. The mean survival of the human fetal RPE xenograft was about 10 weeks with local immunosuppression compared to 4 weeks without Cyclosporin A [67]. In order to evaluate long-term effect of the transplant into the retina, Aramant and Seiler grafted a fetal sheet containing both RPE and neural retina into an immunocompromised model, the nude rat [68]. In that context, all xenografts survived an average of 40 weeks, indicating that the transplant can achieve long survival periods.

Fetal allogenic RPE transplantation showed no visual improvement and signs of rejection in nonimmunosuppressed AMD patients [69, 70]. Another group grafted a suspension of healthy fetal RPE in a 65-year-old legally blind women without systemic immunosuppressant. The patient did not show any visual improvement [71]. In that case, signs of immune rejection were also observed. Radtke and collaborators used intact sheets of fetal neural retina with RPE for grafting into five patients with RP without immune suppression [72]. There was no sign of rejection at 6 months despite HLA mismatch of the patient with the fetal tissue. However, no visual improvement was reported using multifocal ERG, probably due to advanced retinal degeneration. Following these results, a phase II clinical trial was conducted in ten patients (six RP and four AMD) between 2002 and 2005 [73]. Despite no immune suppression, investigators did not report any signs of rejection at 1 year post surgery. However, the pigmentation had decreased over time and was lost in eight out of ten patients at 3 months post operation, and there was no control group. They reported improvement in seven out of ten patients in the ETDRS scale but no improvement using other measures like multifocal ERG [73].

3.2.2 RPE Cells from Adult Eyes as a Source for Cell Therapy

Several cell sources were evoked for RPE cell therapy rising from the adult eye: autologous RPE cells harvested from the peripheral retinal, allogenic RPE graft harvested from cadavers, and recently a new potential source has been identified with the discovery of human RPE stem cells (RPESC).

The group of Suzanne Binder has proposed autologous RPE cell sources. Indeed they demonstrated in 14 human eye cadavers that RPE can be gently harvested and aspirated from the nasal area with a constant efficacy and a sufficient number [74]. The functionality of such cells was previously demonstrated in the RCS rat [75]. A research group of the University of Rochester has used human eye cadavers of 10- and 49-year-old for the isolation of adult RPE cells. Then these cells were grafted in 4-week-old immune-suppressed RCS rats (a time where photoreceptors are still present) [75]. After 1 month, rats were sacrificed and histological evaluation revealed that photoreceptors were preserved. Interestingly RPE from a 49-year-old donor was as effective as the one from a 10-year-old donor in rescuing rat photoreceptors [75]. Phillips and collaborators demonstrated in rabbit the feasibility of harvesting RPE in a subretinal nasal localization and graft the harvested RPE cells to another localization where the RPE cells were previously mechanically debrided [76]. At 30 days post surgery, debrided area grafted with autologous RPE decreases choriocapillaris atrophy and photoreceptors loss compared to debrided areas with no graft. In a prospective pilot study, 13 patients with AMD and foveal choroidal neovascularization (CNV) were operated in order to remove this foveal CNV and were grafted at the same time with autologous adult RPE freshly harvested from the nasal subretinal area [77]. Visual outcomes were encouraging after 17 months with visual acuity improvement in eight eyes without major adverse effects, and a prospective trial with a larger cohort was analyzed. Over a 36 months period, visual gain remained limited, probably due to the poor prognosis expected from these patients [74].

The human adult RPE cell line ARPE-19 was derived from a 19-year-old man who died from a head trauma following motor vehicle accident in 1986. This cell line is easy to grow and retains characteristics of RPE cells through the expression of specific markers and the ability to form cobblestone monolayers [78]. These cells were extensively used to characterize human RPE functions and were used for proof of concept of the cell transplantation approach in the RCS rat. Indeed Pinilla and collaborators had grafted ARPE-19 cells at postnatal day 22 (P22) [79]. Retinas had a preserved connectivity, and photoreceptors were preserved by the treatment. Moreover, ERG recording later during the follow-up (at P90 and P120) demonstrated that transplanted rats maintain their ERG responses compared to control condition, indicating that photoreceptors were still functional [79]. In addition Coffey and collaborators have developed a system to analyze cortically mediated vision following ARPE-19 cell transplantation in RCS rats [80]. In this study, they demonstrated that the primary visual cortex of transplanted rats was responsive to visual stimulation while the one of nonoperated rats was not at 6 months of age.

These different studies demonstrated the feasibility of the use of human adult RPE cells harvested from cadavers to prevent photoreceptors from death. The major limitation for the use of allogeneic RPE from cadavers and autologous RPE is the yield of cells obtained, limiting thus large-scale use due to low amplification potential. For the millions of patients to be treated, a more industrialized manufacturing process is required. To achieve this aim, other cell sources should be investigated.

Recently a new potential source of human RPE came from the discovery of human RPE stem cells (RPESCs) [81]. It was known that RPE cells retain some plasticity like in amphibians where RPE can regenerate retinal cells in vivo upon injury [82]. However, this is not the case in vivo for human RPE cells. In 2012, Salero and collaborators identified a subpopulation of human RPE cells (about 2–3%) that have the potential to be activated upon culture in vitro, acquiring self-renewing properties that they named RPESCs [81]. These cells were harvested from human eyes of 22- to 99-year-old cadavers. They highly express c-Myc and Klf4 indicating stem cell-like properties. RPESC multipotentiality was illustrated through the ability to differentiate into neural or mesenchymal lineages but not into the endoderm lineage. RPESCs are cultured clonally at low density and form non-adherent spheres in a medium containing knockout serum replacement (KSR) supplemented with FGF2 (Fibroblast Growth Factor). Finally, these cells, if they are proven to rescue classical animal models of RPE pathology, might be another potential source of RPE cells for cell therapy. In 2014, Stanzel and collaborators brought the first evidence that these RPE cells obtained from human RPESCs retain fundamental characteristics of RPE when grafted as a monolayer into the rabbit eye [83]. Moreover, these cells survive (95%) for at least 1 month in this model and maintain their monolayer and polarity (ezrin protein localized at the apical side). However, their isolation and amplification remain difficult, and alternative cell types might be better choices.

3.2.3 Alternative Cell Types

The use of functional RPE cells to replace the defective ones is the best treatment to give to patients. However, the sources for RPE cells (essentially fetal or adult RPE sources) were still limited at the beginning of the twenty-first century. The cell transplantation, independently of the cell type, produces a trophic effect that is beneficial for photoreceptor survival. Moreover, some RPE functions like phagocytosis are not restricted to RPE cells. That is why, it is not surprising that several laboratories have developed cell-based therapeutics with different cell types.

3.2.3.1 Iris Pigment Epithelium

RPE and Iris pigment epithelium (IPE) share a common embryonic origin. IPE cells possess phagocytic properties, as demonstrated by the ability of porcine IPE cells to phagocytize rod outer segments [84]. IPE cells form a monolayer with tight junctions suitable to form a de novo blood–retinal barrier. Rat IPE cells were isolated from 20- to 26-day-old and grafted into RCS rats through a transcleral route [85]. At 2 months post surgery, rat IPE was localized between host RPE and retina. They contained phagosomes from rod outer segment uptake. Photoreceptors were found preserved near IPE cells and at some distance from them in transplanted eyes. This indicates that phagocytosis uptake is not the only rescuing mechanism, and trophic factors are probably also released [85]. Another study by Thumann et al. confirmed these results with both transplantation of human and rat IPE in RCS rats [86]. Photoreceptors were preserved at 3 months post surgery to a similar level of ARPE-19 cell-transplanted eyes.

A strategy of autologous human IPE transplantation was tested in 2000 [87]. Human IPE are relatively easy to obtain from patients with a local anesthesia in the same eye than the one to be transplanted. IPE were dissociated and cultured with 15% of autologous serum for 1–2 months. Seven AMD patients with neovascular membranes from 49 to 85 years old were transplanted subretinally. Neovascular membranes were removed through retinotomy just before the transplantation of an IPE cell suspension. Fifteen control patients underwent only neovascularization removal [87]. The majority of operated patients improved their best-corrected visual acuity by two lines or more; however, no statistical difference was found between control and IPE-transplanted patients at up to 13 months of follow-up. The same group followed 56 consecutive patients for at least 2 years using the same procedure [88]. They reported also a statistically improved visual acuity until at least 48 months compared to pretransplantation.

Semkova and collaborators proposed to engraft IPE genetically modified to overexpress the PEDF. The idea was that PEDF would act as an antiangiogenic and a neuroprotective agent to potentiate the effect of transplanted IPE [89]. Indeed, rat IPE was transduced with a high-capacity adenovirus vector having both low toxicity and low immunogenicity. Thereafter RCS rats were transplanted with these IPE cells. The authors found that, at 2 months post surgery, photoreceptors in IPE expressing PEDF-transplanted eyes were better preserved when compared to IPE-transplanted and nonoperated eyes. In addition in a rat model of laser-induced CNV, IPE overexpressing PEDF prevented the formation of new vessels within the laser burns [89]. Another strategy of growth factor overexpression was proposed based on the effect of Brain-Derived Neurotrophic Factor (BDNF) [90]. IPE was transduced with an adeno-associated virus 2 (AAV2) containing the BDNF gene and grafted in a model of phototoxicity. Rats were exposed to constant light for 3 months after the transplantation. The measurement of the outer nuclear layer thickness demonstrated that photoreceptors were preserved in eyes transplanted with IPE overexpressing BDNF compared to normal IPE [90].

The concern with this approach, beside no proven efficacy in human, is the comparability with different batches of IPE from different donors. This strategy is not compatible with the industrial process required to treat at large-scale the millions of patients.

3.2.3.2 Schwann Cells

The Schwann cells produce trophic factors necessary for the survival of photoreceptors, including basic fibroblast growth factor (bFGF), ciliary neurotrophic factor (CNTF), and BDNF. Lawrence and collaborators proposed to graft Schwann cells into the SRS of RCS rats in order to compensate for defective RPE cells. These grafted cells are supposed to provide support for both RPE and photoreceptors improving thus their functionality and their survival [91]. In this study, the supply of the Schwann cells came from the sciatic nerves of congenic neonatal rats. When transplanted in RCS rats, the Schwann cells allowed photoreceptors to survive around the site of injection. Transplanted animals were then analyzed 2 and 3 months after surgery for their visual performances. Using the optokinetic paradigm, the authors demonstrated that transplanted rats were more prone to detect moving stripes compared to control animals up to 3 months post surgery. Keegan and collaborators showed that Schwann cells transplanted in the SRS of rhodopsin knockout ($rho^{-/-}$) mice; another model of acute retinal degeneration preserved the photoreceptor survival until 1 month [92]. They addressed the release of specific factors using reverse transcriptase polymerase chain reaction (RT-PCR) assay for precise measurement of RNA production. BDNF, CNTF, and glial-derived neurotrophic factor (GDNF) transcripts were found, but there was no evidence of NGF and bFGF in Schwann cell culture. The same group demonstrated later that Schwann cells engineered to express GNDF or BNDF improve the survival and function of photoreceptors, when transplanted into the SRS of RCS rat, compared to the same cells not transfected [93]. This highlights the role of these two trophic factors secreted by Schwann cells. For a cell therapy strategy, the supply of Schwann cells could originate directly from the recipient's own cells (e.g., from the sural nerve). This approach might reduce the risk of disease transfer and of immune response.

3.2.3.3 Fetal Brain-Derived Neural Progenitors

Human cortical progenitor cells were proposed and tested in vivo in animal models of RP [94, 95]. Here the idea was not to replace defective RPE or even photoreceptors. Human cortical progenitors are rather expected to produce trophic factors that might preserve photoreceptors. This is the mode of action of such cells when grafted in neurodegenerative disease models [94]. The human fetal cortex could be obtained from a brain tissue at 21 weeks of gestation and cultured as neurospheres for several passages [95]. When such cells were grafted into the SRS of RCS rats, they showed the potential to sustain the vision of RCS rats for a long period of time (280 days) as evaluated by the optokinetic paradigm. Interestingly, histological analysis revealed that grafted cells formed an "RPE-like layer" into the SRS, with some cells being pigmented [94]. However, they did not express typical RPE markers, and the cells were still proliferating into the SRS until at least P150 [95]. The human cortical progenitors cells were also injected into the SRS of six normal rhesus monkeys [96]. In that context, cells were transduced with Green Fluorescent Protein (GFP) to follow them. Monkeys recovered well from the surgical procedure, and there was no disturbance in multifocal ERG. Cells survived as a monolayer at 39 days post operation even if treated only with 5 days of topical steroids [96]. The authors concluded that the surgery might be safe for human, and that the cells should survive without need of systemic immunosuppression.

Human central nervous system stem cells (HuCNS-SCs) were purified by StemCells, Inc. in 2000 [97]. A single fetal brain tissue (16–20 weeks of gestation) was exposed to monoclonal antibodies for cell-surface markers and cell sorted. The HuCNS-SC express high levels of CD133 and low levels of CD24. In addition, they do not express antigens of the hematopoietic lineage CD45 or CD34. These cells can self-renew, form neurospheres, and can be banked. They are developed for the treatment of central nervous system diseases and were tested for retinal protection also in AMD. When HuCNS-SCs were transplanted into the SRS of the RCS rat, photoreceptors were protected and the vision was preserved [98]. Donor cells remain immature and do not express retinal markers during at least 7 months. In addition, the authors reported limited proliferation. StemCells, Inc. launched an ongoing phase I/II clinical trial in US in 16 patients with geographic atrophy secondary to AMD in order to investigate safety and preliminary efficacy [99].

3.2.3.4   Bone Marrow Mesenchymal Stem Cells

Bone marrow mesenchymal stem cells (MSCs) are precursors of bone, cartilage, and adipocytes in vivo, and they provide a suitable environment for the hematopoietic stem cell niche [100]. These cells have the major advantage to be harvested from bone marrow and expanded, thus allowing autologous transplantation. They are also poorly immunogenic, and it was demonstrated that MSCs secrete trophic factors that promote neuronal survival, like BDNF or bFGF [101]. Arnhold and collaborators tested the potential of MSCs for the rescue of retinal degeneration in RCS rats [102]. The authors used rat MSCs transduced with GFP in order to follow their fate in the rat eye. Transplanted MSCs adopted, in that context, an RPE-like morphology at 2 months post surgery. This was also demonstrated in vitro by cocultures of MSCs and RPE, where MSCs adopted a typical RPE morphology (intracellular pigment granules, hexagonal morphology, and cytokeratin expression). Photoreceptors of transplanted retinas were still numerous compared to the control ones at 2 months post operation [102]. Inoue and collaborators transplanted mouse MSCs into the SRS of RCS rats. It resulted in the slowdown of retinal degeneration with a drastic reduction of photoreceptor loss at

8 weeks after surgery [103]. In addition, the authors analyzed retinal functions and found that transplanted animals had a better ERG preservation compared to SHAM-treated eyes at both 4 and 8 weeks post operation. As a conditioned MSCs medium is able to promote survival of mouse retinal cultures, the authors suggested that the rescue effect observed in RCS rats might be mediated by the release of trophic factors that improve the environment [103]. In this study, MSCs remained in the SRS without any migration to the retinal layers. By contrast to Arnhold and collaborators, the authors did not report any signs of transdifferentiation of MSCs into RPE-like cells. Human adult bone marrow-derived somatic cells (derived using a proprietary process) were finally tested in RCS rats in 2010 [104]. The authors showed that transplanted RCS rats had a better visual acuity in the optokinetic test up to 3 months postnatal despite the absence of cell at that time. The survival of cells was not improved with a cyclosporine A treatment. Another report stated that human bone marrow MSCs grafted into the SRS of RCS rats did not survive more than 2 weeks [105]. They demonstrated also that the effects were sustained for 20 weeks, and that a second transplantation did not improve the phenotype.

3.2.3.5 Retinal Neurospheres

In the adult mammalian eye, neural progenitors are located in the pigment ciliary bodies [106]. These cells are rare, quiescent but can be stimulated to proliferate in vitro with bFGF. They can be amplified as neurospheres, and they are named retinal neurospheres (RNS) [107]. RNS retain their pigmentation and express markers like the neuroectodermal marker nestin. They are multipotent as they give rise to neurons, astrocytes, and oligodendrocytes. Rat or mouse RNS express also Chx10, a homeobox gene that is a retinal progenitor marker [106, 108]. In human, RNS were derived from the pars plicata and pars plana of the retinal ciliary margin [109]. These cells were isolated from early postnatal to 70-year-old postmortem donors. When cultured under specific differentiation media, they are able to give rise to different retinal cell types and to RPE. Human RNS were transplanted into immunodeficient NOD/SCID mice of 1-day-old. The authors showed that the cells survived at 28 days and gave rise to retinal progeny. RNS stem cell nature is controversial; some scientists stated that the pigmented ciliary epithelium contains differentiated cells that ectopically express markers upon specific culture conditions [110, 111]. Finally, RNS cells should be tested into animal models of retinal degeneration to demonstrate their therapeutic potential.

3.2.3.6 Umbilical-Cord Stem Cells

Human umbilical tissue-derived cells (hUTCs) are a promising cell source as it can be amplified to at least 1×10^{17} cells from a single donor without genetic abnormalities or changes in phenotype [112]. Cells are obtained by a simple enzymatic digestion of the human umbilical cord following normal births. When grafted into the SRS of RCS rats, hUTCs improved ERG responses to light compared to SHAM-treated animals. In addition, visual acuity evaluated by the optokinetic response to moving stripes was preserved in hUTCs-treated rats at 2.5 months post surgery [112]. Neurotrophins secreted by hUTCs, like BNDF, are suggested to mediate the preservation of photoreceptors.

3.3   RPE Derived from Pluripotent Stem Cells

Human embryonic stem cells (hESCs) and human-induced pluripotent stem cells (hiPSCs) represent the two principal types of pluripotent cells. These unspecialized cells are able to self-renew and could give rise to all somatic cell types. Compared to the other cell sources described earlier, banking of these cells might be up scaled at an industrial level to treat millions of AMD and RP patients.

In 1998, the first hESC lines were derived from the blastocyst inner cell mass of human embryos produced by in vitro fertilization. These hESC lines showed unlimited proliferation ability in vitro and maintained the potential to form derivatives of the three embryonic germ layers: endoderm, mesoderm, and ectoderm [113].

The reprogramming of adult somatic cells, into artificial pluripotent stem cells, called induced pluripotent stem cells (iPSCs) by Yamanaka lab allowed to overcome ethical constraints associated to the use of human embryos [114]. Indeed human adult somatic cells, typically human fibroblasts, were forced to express pluripotency genes by retroviral transduction of four transcription factors: Oct3/4, Klf4, Sox2, and c-Myc [114]. This was reproduced by Thomson's lab using a different cocktail of factors [115].

The hiPSCs are comparable to hESCs in terms of morphology, cell growth, surface markers, gene expression, in vitro differentiation, and teratoma formation when transplanted into immunocompromised animals [114]. The potential to generate autologous cell sources from reprogrammed patient's somatic cells make hiPSCs very attractive for clinical applications [116]. These two cell sources (hESCs and hIPSCs) were used to produce RPE cells that might treat RPE-associated degenerative eye diseases.

3.3.1   Clinically Compatible Protocols to Obtain and Bank RPE Cells

Significant research efforts have focused on finding the ideal method for efficiently deriving RPE from embryonic and induced pluripotent stem cells. It has been well demonstrated that hPSCs have the potency to "spontaneously" differentiate into RPE using the continuous adherent culture method or the embryoid body method [117].

Protocols published in the literature could be gathered in three main methods: spontaneous differentiation, intermediate embryoid body (EB) formation, and 3-D cellular aggregates. In recent clinical trials, the first two types were used in order to generate GMP (Good Manufacturing Process)-compatible RPE cells. In 2004, Klimanskaya and collaborators reported the first spontaneous differentiation of hESCs into RPE cells [118]. To achieve this differentiation, the hESCs were cultured on mouse feeder cells in absence of FGF2. After 6–8 weeks, the authors observed the presence of brown-pigmented regions in over-confluent cultures. Pigmented regions were isolated and subcultured to generate the RPE cell population. The limitation with this protocol was the low efficiency regarding the number of cells obtained and the requirement of the manual selection of pigmented areas.

To improve the efficacy of RPE differentiation protocols, several teams developed protocols based on developmental studies made in xenopus, zebrafish, and mouse. This approach has led to the use of an increasing number of growth factor and small-molecule supplements in the culture media [117]. The pathways targeted vary from protocol to protocol, but the inhibition of TGF-ß pathway using noggin, LDN-193189, and/or SB-431542 is commonly used to engage neuroectodermal differentiation [119]. Neuroectodermal progenitors are then specified toward an anterior fate using inhibitors of the WNT signaling pathway (Dkk1/Nodal and Lefty A) to obtain eye field progenitors [120]. Using this method, Lamba's team showed that it was possible to obtain nearly 80% of retinal progenitors [121]. For now, one of the quicker and the more efficient protocol was published by Buchholz and collaborators [122]. This team combined Lamba's protocol with already described RPE-inducing factors such as nicotinamide and activin A in order to obtain 78% of RPE cells at day 14 [122]. One year later, the same team published that the RPE differentiation is even better with the use of activator of the WNT pathway such as CHIR99021 at the end of the differentiation process [123]. RPE cells could be obtained without these cytokines, but the efficiency is much lower [124]. Whatever the protocol used, the RPE obtained from pluripotent stem cells are similar to primary human fetal RPE [125]. Recently, a quantitative real-time PCR-based high-throughput screen was used to establish a new differentiation protocol with small-molecule treatment and that is suitable for a clinical application. Using this strategy, Chetomin, an inhibitor of hypoxia-inducible factors, was found to strongly increase RPE differentiation. The combination with nicotinamide resulted in conversion of over one-half of the differentiating cells into RPE. Single passage of the whole culture yielded a highly pure hPSC-RPE cell population that displayed many of the morphological, molecular, and functional characteristics of native RPE [126].

The second strategy to generate RPE cells involves an intermediate step: the embryoid body (EB) formation. The pluripotent stem cells, when cultured onto non-adherent culture dishes form aggregated structures that resemble early postimplantation embryos and that are therefore called EBs [127]. Pluripotent stem cells derived in EB structures acquire markers specific to the three embryonic germ layers [128] and could differentiate into multiple cell lineages. To induce the neuroectodermal lineage, the EBs obtained are then cultured in adherence, and after 4 additional weeks, pigmentation starts rising. These pigmented regions are expanded and ultimately form a monolayer of hexagonal pigmented cells [129]. This kind of differentiation can also be improved by the sequential use of nicotinamide and activin A, although the mechanism by which nicotinamide improves neural and eye field differentiation remains elusive [130].

In the last years, many research teams have explored new approaches to increase further the efficacy of RPE differentiation protocols. The most innovating approaches are represented by the generation of RPE cells from tridimensional structures. Gamm and collaborators developed a protocol where the pluripotent stem cells are differentiated into neural rosette, a primitive anterior neuroepithelium structure [131]. The neural rosette structures were isolated and cultured in suspension. After 20 days of culture, 3-D cell aggregates appeared and formed an optical vesicle-like (OV) shape. These tridimensional structures contained photoreceptors and RPE cell progenitors. To generate RPE cells, OV-like structures were cultured in presence of activin A for 20 days. Pigmented OV-like structures were plated onto coated dishes in order to promote RPE cell proliferation. This protocol was used to generate RPE cells from hESCs and hiPSCs in a reproducible manner [131]. Recently, Goureau and collaborators developed a new protocol of self-forming neural retina structures containing RPE and retinal progenitor cells including precursors of photoreceptors, without addition of exogenous molecules [132, 133]. In this study, hiPSCs were cultured in a serum-free medium containing N2 supplement. To generate RPE cells, the FGF was removed from culture medium to promote the neuroectoderm induction. After 14 days, neuroepithelial-like structures with pigmented areas were observed. These pigmented patches were mechanically isolated and cultured on gelatin-coated dishes. The pigmented patches proliferated and formed the typical cobblestone epithelial structure of RPE [133].

Recently, automated retinal differentiation including RPE differentiation was reported by Duncan E. Crombie and collaborators [134]. They showed the ability to maintain and subculture pluripotent stem cells using a customized automated platform. Then, they differentiated these cells into RPE cells using a directed protocol and reported the apparition of pigmented areas at day 35. Although, this is a promising step toward the scale up of the RPE differentiation that will be required for future applications; further works are necessary to obtain a pure population of cells with this method [134].

For the clinical translation step of RPE cell differentiation protocols, all parameters need to be evaluated with utmost rigor. According to European and US regulatory guidelines, the production of RPE cells requires to be reproducible and to use a fully defined medium. Thus, the constituents of differentiation and culture medium, cytokines and the matrix used to coat cell dishes, are selected based on their origin; their certificates indicating that they are pathogen-free and produced in a sterile environment. Moreover, the manufacturer should provide a full traceability of their starting and raw materials. All this selection increases the translation from the initial research protocol to the first clinical trial. This optimization process faces challenges due to the risk of inconsistency between the non-GMP starting/raw material and the GMP one and due to the batch to batch variability of a same product.

In 2012, Astellas Pharma (previously named OCATA Therapeutics) announced the first preliminary data of the clinical application of RPE derived from hESCs [135]. In this phase I/II clinical trial, a hESC line derived from in vitro fertilization embryo named MA09 was used. The MA09 cell line was cultured and amplified on Mitomycin C-treated mouse embryonic fibroblast (MEF) in GMP facility to generate a clinical Master Cell Bank. This manufacturing process used the EB formation process to generate RPE cells. hESCs were treated with 0.05% trypsin-EDTA to detach from feeder cells and subsequently plated into low-attachment cell dishes to promote the EB formation in MEM medium containing the B27 supplement. The EBs were maintained in culture for 1 week and seeded on gelatin-coated plates until the pigmented colonies appeared. Using enzymatic and mechanical methods, the pigmented RPE cells were isolated and cultured in Endothelial Growth Basal Medium on gelatin-coated cell dishes. The RPE cells were amplified for two passages, before being frozen to create the RPE clinical bank. The hESC-derived RPE cells were thawed and injected as a cell suspension into the SRS of patients suffering from dry AMD and Stargardt's disease. The RPE cryovials could be sent frozen to different hospitals without the requirement of specific cell culture materials at the site of surgery. This allows a large-scale use of such cell product.

A second clinical trial using hPSCs-derived RPE for retinal degenerative diseases started in September 2014 in Japan. This study represents the first clinical assay using autologous iPSC therapy for the exudative form of AMD. In their study, Dr. Takahashi and her team have transplanted sheets of hiPSCs-derived RPE cells without any artificial matrix or scaffold ([116, 136]; Press Release, RIKEN July 30, 2013). hiPSCs were cultured and amplified on mouse embryonic fibroblast (MEF) feeder cells. To generate RPE cells from hiPSCs in xeno-free conditions, a direct differentiation protocol was used, in which there was no intermediate stage of EB formation. That process used two differentiation media and two different coatings. The undifferentiated hiPSCs were cultured in feeder-free conditions using gelatin-coated plates. During the differentiation process, the cells were cultured with Glasgow Minimum Essential Medium (GMEM) in presence of decreasing percentage of Knockout Serum Replacement (KSR) (from 20% to 10% during 20 days of culture). To induce RPE differentiation, some molecules were added in the culture medium during the first 18 days like ROCK inhibitor (Y-27632), a TGFβ inhibitor (SB 431542), and a protein kinase inhibitor (CK17) [137]. After 3 weeks of culture, pigmented cells appeared and the differentiation media was replaced with DMEM/F12 supplemented with B27 [116].

Our lab, in collaboration with the Institut de La Vision (Paris, France) is developing another protocol based on hESCs using a spontaneous differentiation protocol without growth factors or small molecules [138, 139]. The clinical grade hESC line, named RC-09 (Roslin Cell Laboratory, Edinburgh), was amplified in feeder-free conditions, in a GMP facility, to create a clinical hESC Master Bank. To differentiate the hESCs into RPE cells, the RC-09 are cultured in DMEM medium supplemented with 10% of KSR in the absence of FGF2. After 4–5 weeks of cultures, the presence of pigmented regions in the overgrowth colonies of RC-9 could be observed. These pigmented cluster cells are then mechanical isolated and cultured in culture dishes in presence of 4% of KSR (Passage P0). The RPE cells are amplified and frozen at passage P1, to create the RPE Bank.

All these culture methods will be useful for the demonstration of the safety and efficacy of hPSCs-derived RPE in ongoing and future clinical trials. The next challenges will be the development of commercially viable sources of RPE that are low cost, reliable, and robust.

3.3.2 Quality Controls of Differentiated RPE Cells

In most countries, the use of cellular products for medical use is regulated by governmental agencies to ensure the protection of patients, so that novel therapies will be the most widely beneficial for the population.

Many technical and regulatory breakthroughs in the last few years have made stem cell-based treatment for retinal degeneration more plausible. During the manufacturing process of hPSCs-derived RPE cells, several important limiting key points need to be carefully evaluated according to regulatory guidelines. It includes safety, purity, identity, potency, and stability of differentiated cells [135, 140]. Even minimal manipulation of the cells outside the human body introduces a risk of contamination with pathogens, and prolonged passages in cell culture carries the potential for genomic and epigenetic instabilities that could lead to cell dysfunctions or frank malignancy.

The purpose of preclinical studies is to (1) provide evidence of product safety and (2) establish proof of principle for therapeutic effects. International research ethic policies, such as the Declaration of Helsinki and the Nuremberg Code, strongly encourage the performance of animal studies prior to clinical trials. Before initiating clinical studies with stem cells in humans, researchers should have persuasive evidence of clinical promise in appropriate in vitro and/or animal models.

3.3.2.1 Safety of Therapeutic Cells

All reagents should be subjected to quality control systems to ensure the quality of the reagents (raw and starting materials) prior to introduction into the manufacturing process. For extensively manipulated stem cells intended for clinical application, GMP procedures should be strictly followed [141].

The RPE cells should be checked for safety, RPE identity, potency, and functional activity at various steps of the manufacturing process (in-process testing during the manufacturing, after RPE thawing and in the final clinical product) [135]. The absence of bacterial and mycoplasma contamination all along the process and in the final product in its final formulation should also be tested. Components of animal origin may present a risk of transferring pathogens or unwanted biological material. For example, when cultured onto MEF (Mouse embryonic Fibroblasts), the cells should also be monitored for the absence of residual murine viruses. In some circumstances, it may not be possible or optimal to substitute these components. Researchers need to demonstrate that the risk to use such product is mitigated by the different additional safety tests of these animal-based components.

hPSCs, regardless of particular cell type, carry additional risks due to their pluripotency. These include the ability to acquire mutations when maintained for prolonged periods in culture, to grow and differentiate into inappropriate cellular phenotypes, to form benign teratomas or malignant outgrowths, and to fail to mature in the cell type of interest. Teratomas are tumors of multiple lineages, containing tissues derived from the three germ layers [142]. It confers additional risk to patients, and appropriate tests must be planned to ensure safety of stem cell-derived products [143]. When undifferentiated, hPSCs are transplanted into immunocompromised animals; they might give rise to these teratomas. For hPSCs-derived products, a quantification strategy needs to be developed to evaluate the risk of remaining undifferentiated cells in the final product. The level of impurities in the final product should be enough low to prevent teratoma formation in long-term animal studies (over 6 months). Tumorigenicity assays should actually include groups of animals transplanted with undifferentiated hPSCs and other groups injected with serial dilutions (spiking studies) of undifferentiated hPSCs with hPSCs-derived RPE cells to set a contamination limit that is susceptible to induce a teratoma and that should not be reached in the cell therapy product. The cells used in tumorigenic test (regulatory toxicity studies) should be produced with a manufacturing process similar to the one used for the clinical application [143]. Several immunodeficient animal models are now available for such studies: NOD/SCID (NOG) mice, and NOD/SCID/IL-2rg KO (NSG) mice, Rag2-yC double-knockout (DKO) mice. These animal models are T-cell, B-cell, and NK-cell-deficient and show very high engraftment potential of human cells compared to nude mice (T-cell-defective) [142].

Other important issues to consider in safety assays are related to the biodistribution of grafted cells. The stem cells and their differentiation derivatives may have the potential to migrate from the injection site to other organs. Careful evaluation of biodistribution, assisted by sensitive techniques of imaging and monitoring of homing, retention, and subsequent migration of transplanted cell populations, is crucial for measuring cell dispersion. While rodents or other small animal models are typically a necessary step in the development of stem cell-based therapies, they are likely to reveal only major toxic events [142]. Biodistribution tests are performed in injured or healthy animals using the route of administration of the cell therapy product. The proximity of many physiological functions between large mammals and humans may favor testing the biodistribution and toxicity of a novel cell therapy in at least one large animal model. Additional histological analyses or banking of organs for such analysis at late time points is recommended. Depending on the laws and regulations of the specific country, biodistribution and toxicity studies often need to be performed in a GLP (Good Laboratory Practice)-certified animal facility.

In Astellas Pharma preclinical toxicity studies, the hESC-derived RPE was injected in the SRS of NIH III mice, a typical nude mice characterized by the absence of thymus and T-cell function and by complications in the maturation of

T-independent B cells. The absence of human cells in the eyes and others organs of these mice was determined by RT-qPCR and immunostaining at 4, 12, and 40 weeks after grafting. All tests showed no safety issues in any animal even if hESCs-derived RPE cells were spiked with 0.01% of undifferentiated cells [135].

Similar results were obtained with hiPSCs-derived RPE cell sheets transplanted into the SRS. Kamao and collaborators did not observe any tumor formation in various animal models (RCS rat, immunodeficient mouse and monkey) [116, 136].

Simplified culture conditions are essential for large-scale drug screening and medical applications of hPSCs. However, hPSCs are prone to genomic instability, a phenomenon that is highly influenced by the culture conditions (various culture media, passaging techniques, and culture feeders/matrices) that may differently affect genetic stability or impart different selective pressure on cells [144, 145]. For example, enzymatic dissociation, a cornerstone of large-scale hPSCs culture systems, is deleterious, but the extent and the timeline of the genomic alterations induced by this passaging technique are still unclear [144]. For the karyotype analysis of undifferentiated hPSCs and their derivatives, the G-banding and FISH methods are used. In the protocol developed by Astellas Pharma, the chromosomal integrity of hESC Master Cell Bank was evaluated by FISH and G-Banding methods [135]. The reprogramming process of somatic cells into pluripotent stem cells can induce chromosomal abnormalities, as highlighted recently. The Japanese clinical trial team has identified the presence of mutation in the hiPSCs generated from the second patient [136]. Three single-nucleotide variations and three copy-number variants were detectable in the hiPSCs while were absent in the original patient fibroblasts. For this reason, it was decided not to treat the second patient with autologous hiPSCs-derived RPE [146].

3.3.2.2 Purity of the Differentiation Process

To evaluate the purity of hPSCs-derived cells, a panel of RPE-specific markers should be used, like ZO-1 (tight junction marker), MITF, and PAX-6 (eye field specification), RPE-65 and CRALBP (retinoid cycle), Bestrophin (chloride channels), and MERTK (phagocytosis function) [116, 138].

In the protocol developed by Astellas Pharma, the hESC-derived RPE was characterized by immunofluorescence. hESCs-derived RPE cells should coexpress more than 95% of PAX-6 and MITF, 95% of PAX-6 and Bestrophin and express more than 95% of ZO-1 [135]. The purity of RPE cells generated from autologous hiPSCs, in the Takahashi's protocol, was characterized by the quantification of PAX-6 or Bestrophin-positive cells [116, 136].

Gene expression of pluripotency and RPE-specific genes could be monitored by quantitative PCR. The absence of contaminating hPSCs in the RPE population might be evaluated by the downregulation of OCT-4, NANOG, and SOX-2 mRNA expression compared to undifferentiated pluripotent stem cells. The monitoring of the upregulation of some specific RPE markers, like RPE-65, PAX-6, CRALBP, and BEST1, indicates the RPE identity [116, 136, 138].

3.3.2.3 Functionality of the RPE Obtained from Differentiation

The functional assessment is based on typical characteristics of RPE cells as the phagocytosis activity, the polarized secretion of growth factors, and the barrier function of epithelial cells. To maintain the excitability of photoreceptors, POS undergo a constant renewal process [6, 31]. In the protocol developed by Astellas Pharma, the phagocytosis activity of hESCs-derived RPE, was monitored through the uptake of fluorescent bioparticles. The internalization of bioparticles is measured in this approach by flow cytometry [135]. Other systems use porcine POS labeled with fluorescent markers. When RPE cells are exposed to such POS, they are able to phagocyte them [133, 139].

Maintaining the polarized secretion of growth factors is a crucial parameter for restoring some RPE functions after the graft. Two growth factors are particularly important: PEDF (apical secretion) essential for the structural integrity of the retina, and the VEGF (basal secretion) essential for the integrity of choriocapillaries. The concentrations of PEDF and VEGF in a conditioned RPE medium could be measured by ELISA assays and reflect functional properties of RPE cells [116].

The measure of RPE epithelial resistance is an additional method to evaluate the potency of hPSCs-derived RPE cells. Tight junctions are a type of cell–cell adhesion that has a fundamental role in the functionality of the RPE layer [147]. Tight junctions form a partially occluding seal that retards diffusion of solutes across the paracellular space. This barrier permits the RPE layer to establish and maintain concentration gradients between its apical and basal environments [148]. The strength of this barrier may be measured by the transepithelial electrical resistance in vitro [136].

3.3.3 Injection Strategy of the Cell Therapy Product

Significant research efforts are focusing on finding the ideal method for transplanting hPSCs-derived RPE cells into the subretinal space. Optimizing RPE transplantation procedures resulted in the development of two different therapeutic strategies: (1) introducing a cell suspension of nonpolarized RPE cells into the subretinal space and allowing the donor cells to integrate within the host retina, and (2) transplanting polarized sheets of RPE to allow for improved safety and better clinical outcomes, since normal RPE functions are dependent on specific cellular features of its apical and basal domains [125, 149].

In phase I/II clinical trials sponsored by Astellas Pharma, the RPE cells derived from hESCs were delivered as a cell suspension into SRS of dry AMD and Stargardt's patients [135]. 5×10^4 hESCs-derived RPE cells in 150 μL were injected into a preselected region of the pericentral macula that was not completely damaged [125]. Three cell doses were initially proposed for injections: 5×10^4, 1×10^5, and 1.5×10^5 RPE cells in 150 μL in order to evaluate the safety of the cell therapy. To avoid the dispersion of donor RPE cells in other site than the retina, the injection site was carefully chosen based on the presence of native RPE layer albeit damaged, to improve the transplant integration [140].

The most important point of success for cell transplantation into SRS region is the integrity of the implanted region. Retinal degenerative diseases, like AMD may lead to the destruction of RPE-Brüch's membrane, compromising the integrity of the blood–retinal barrier. Thus, if RPE cells are transplanted as a cell suspension, their chance to survive and integrate correctly might be compromised. To overcome this problem, a RPE monolayer that exhibits the physiology of its natural counterpart represents a favorable alternative to RPE cell suspension [83].

Takahashi and collaborators developed a strategy based on RPE sheets transplantation without any matrix or scaffold using an ingenious system [116]. Indeed, hiPSCs-derived RPE are cultured onto Transwell inserts coated with collagen type I. When the RPE cells reach confluency and form a typical cobblestone pigmented epithelium, the RPE sheet is removed from the insert by a collagenase treatment [116, 136]. The size of the RPE sheet is then adjusted to 1.3 × 3 mm using a laser microdissection system.

We developed an alternative approach based on a natural scaffold to graft an RPE sheet. Our cell therapy product is constituted of a three dimensional (3D) patch of hESCs-derived RPE cultured on a human amniotic membrane (hAM) [138]. hAMs are obtained from cesarean sections of normal births [150]. The hAM consists of an epithelial monolayer, a thick basement membrane, and a multilayer of collagen. The components of the hAM create an interesting native scaffold for cell seeding in tissue engineering. Moreover, the hAM presents significant biological advantages like anti-inflammatory, healing, and antimicrobial properties [151]. For all these reasons, the hAM is commonly used in clinical applications for ocular surface reconstruction [152]. In this approach, the hAM is treated in order to remove the native epithelia cell layer, and the denuded hAM is attached to a culture insert, which allows the culture of RPE cells. The hESCs-derived RPE cultured on this natural scaffold, form a typical cobblestone pigmented epithelial layer, with polarized secretion of VEGF (Fig. 3.2).

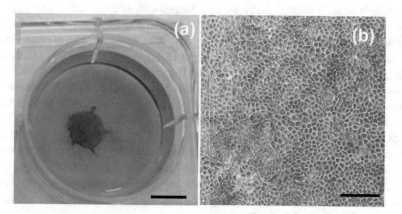

Fig. 3.2 Macroscopic (**a**) and microscopic (**b**) observation of hESCs-derived RPE cultured for 4 weeks on human amniotic membrane. The RPE cells exhibit typical pigmented cobblestone morphology. Scale bar 1 cm (**a**) and 100 μm (**b**)

Synthetic scaffolds may also be used to transplant RPE cells. Several parameters must be taken into consideration, as thickness, mechanical properties, and biodegradation, to prevent additional damage of the retina, and improve the interactions between the retina and RPE [153]. The transplantation of such polarized RPE monolayer on ultrathin parylene substrates has proven its safety and efficacy [154].

Coffey and collaborators developed a porous polyester scaffold who serves as a matrix for the transplantation of the hESCs-derived RPE. The targeted patients for this cell therapy suffer from wet AMD (Press Release UCL, Sept 29, 2015; ClinicalTrials.gov NCT01691261) [155].

3.4 Clinical Trials for the Treatment of AMD and RP

3.4.1 Adult and Fetal RPE

Use of adult and fetal RPE cells demonstrated the potential of the cell therapy approach for treating RPE-associated diseases [74, 156]. However, their reduced availability has limited their potential for large-scale use in clinic. One of the last strategies taking advantage of adult RPE cells is the one of the group of Sally Temple (Neural Stem Cell Institute, NY) who had discovered and isolated RPESCs from adult RPE cell culture [81]. They are now developing cGMP optimization of allogeneic cells from adult RPE in order to target AMD patients with geographic atrophy [157].

3.4.2 hiPSCs/hESCs-Based Clinical Trials

Several phase I/II clinical trials were launched by Astellas Pharma (Table 3.1) using allogeneic hESCs-derived RPE as cell suspension both in USA and UK for Stargardt's macular dystrophy (SMD) and AMD. The same cells were also used by CHABiotech CO., Ltd. in the Republic of Korea to treat patients with AMD. Preliminary reported data indicated no major safety issue [135, 140, 158]. Astellas Pharma is moving forward with a phase II clinical trial, which is about to be launched (NCT02563782). Recently, the London Project to Cure Blindness and Pfizer has started a phase I/II clinical trial with the first wet AMD patient being treated in 2015. Here the transplant is composed of RPE derived from allogeneic hESCs cultured over a polyester membrane [159]. Other clinical trials based on allogeneic hESCs-derived RPE are currently recruiting AMD patients in California (NCT02590692) and Israel (NCT02286089). In France, a consortium composed of I-Stem and Institut de la Vision is currently optimizing a cell therapy with allogeneic hESCs-derived RPE over a biological substrate in order to launch a phase I/II clinical trial targeting RP in 2019 [138].

Table 3.1 Clinical trial planned or ongoing based on hESCs or iPSCs in order to treat RP and/or AMD

Stage of development	Targeted disease	Sponsor/company	Therapy
Phase II NCT02563782	AMD	Astellas Pharma	Allogeneic hESC-RPE, cell suspension injection
Phase I/II NCT01469832	SMD	Astellas Pharma	Allogeneic hESC-RPE, cell suspension injection
Phase I/II NCT01344993	Dry AMD	Astellas Pharma	Allogeneic hESC-RPE, cell suspension injection
Phase I/II NCT02122159	MMD	Astellas Pharma	Allogeneic hESC-RPE, cell suspension injection
Phase I/II NCT01345006	SMD	Astellas Pharma	Allogeneic hESC-RPE, cell suspension injection
Phase I/II NCT01674829	Dry AMD	CHABiotech CO., Ltd	Allogeneic hESC-RPE, cell suspension injection
Phase I/II NCT02286089	AMD	Cell Cure Neurosciences Ltd.	Allogeneic hESC-RPE, cell suspension injection
Phase I/II NCT02590692	Dry AMD, GA	Regenerative Patch Technologies, LLC	Allogeneic hESC-RPE, epithelium on a parylene membrane
Phase I/II	AMD	RIKEN Centre for Developpmental Biology	Autologous hiPSC-RPE, epithelium without substrate
Phase I NCT01691261	Wet AMD	Pfizer/London Project to Cure Blindness	Allogeneic hESC-RPE, epithelium on polyester membrane
cGMP optimisation	AMD, SMD, RP	Cellular Dynamics International/NEI	Autologous hiPSC-RPE, epithelium on biodegradable scaffold
cGMP optimisation	Dry AMD, RP	I-Stem/Institut de la Vision, France	Allogeneic hESC-RPE, epithelium on amniotic membrane
Preclinical	AMD, Best disease, LC A	H MC, Israel	Allogeneic hESC-RPE, cell suspension injection

When registered in clinicaltrial.gov website, the reference number of the clinical trial is indicated
SMD Stargardt's macular dystrophy, *AMD* age-related macular degeneration, *GA* geographic atrophy, *RP* retinitis pigmentosa, *LCA* Leber's congenital amaurosis, *MMD* myopic macular degeneration

As an alternative to allogeneic cells, other laboratories develop clinical trials based on autologous hiPSCs. The Riken Center for developmental biology in Japan has initiated in the end of 2014 the first phase I/II clinical trial based on autologous hiPSCs derived into RPE. The first patient had a wet AMD with CNV, and five other patients were planned. However, due to genetic defects found in the cells from the second patient, this first-in-man clinical trial was suspended [136, 146]. A Japan's new law facilitates now the commercialization of hiPSCs. Indeed, regenerative therapy will be conditionally approved if they are demonstrated safe. Then they have up to 7 years of commercialization to demonstrate efficacy. New strategies have been adopted, and Riken now pushes toward the use of allogeneic hiPSCs to reach earlier the market. The National Eye Institute (NIH) and Cellular Dynamics International are developing autologous hiPSCs-derived RPE cultured on a biodegradable scaffold [157]. They are currently optimizing the cGMP process and expect to treat the first patient in 2018.

3.4.3 Other Stem Cell Trials

Autologous BM-derived stem cells, delivered intravitreally, are developed by many laboratories around the world (Table 3.2). The surgery is easier as a trophic effect is expected; the cells do not need to be located in the SRS like RPE cells.

Human central nervous system stem cells derived from fetal brain are under investigation for dry AMD patients. The phase I/II clinical trial was started in 2012, and preliminary results in 15 patients indicate no safety issues (Table 3.3). The Phase II study is currently recruiting patients. Other companies use human retinal progenitors injected either through an intravitreal (jCyte, Inc.) or subretinal route (ReNeuron Limited). jCyte, Inc. had currently treated 4 out of 16 patients at the 2015 summer in California. They will be followed for 12 months to report any safety issue. The effect of the cells is expected to be trophic.

An ongoing phase I/II clinical trial sponsored by Janssen Research & Development, LLC is based on a subretinal delivery of human umbilical tissue-derived stem cells. The cells are delivered through a catheter delivery system [159]. Finally, a clinical trial was launched based on autologous adipose-derived stem cells (NCT02024269) by the Hollywood Eye Institute in Florida and US Stem Cell, Inc. The cells are obtained via liposuction. After isolation of adipose-derived stem cells, they are injected into the vitreous [160]. However, recent results observed in three patients treated in a stem cell clinic with such cells raised concerns about their safety. Loss of vision acuity was reported, one patient been blind due to the treatment [161].

3.5 Management of the Graft Rejection

The eye is a prototypic immune-privileged tissue that resists immunogenic inflammation through multiple mechanisms [162, 163]. Inside the eye, the SRS is even a better transplantation site than the vitreous cavity. Indeed, cells grafted into the SRS demonstrated better survival than the ones transplanted into the vitreous cavity [105, 164]. Moreover, stem cells have low immunogenic capacities [165–167],

Table 3.2 Clinical trial planned or ongoing based on bone marrow (BM) stem cells in order to treat RP and/or AMD

Stage of development	Targeted disease	Sponsor/ company	Therapy
Phase I NCT01531348	RP	Mahidol University	Intravitreal injection of BM-derived MSCs
Phase I/II NCT02016508	AMD	Al-Azhar University	Intravitreal injection of autologous BM derived stem cells
Phase I/II NCT01518127	AMD and SMD	University of Sao Paulo	Intravitreal injection of autologous BM stem cells
Phase II NCT01560715	RP	University of Sao Paulo	Intravitreal injection of autologous BM stem cells
Phase I NCT01068561	RP	University of Sao Paulo	Intravitreal injection of autologous BM stem cells
Phase I/II NCT01914913	RP	Chaitanya Hospital, Pune	Autologous BM derived Mono Nuclear Stem Cell (BMMNCs)
Phase I NCT01736059	Non-exudative AMD, diabetic retinopathy, retina vein occlusion, RP, hereditary macular degeneration	University of California, Davis	Autologous CD34+ bone marrow stem cells, intravitreal injection
Phase I NCT02280135	RP	Red de Terapia Celular	Intravitreal injection of autologous BM stem cell

When registered in clinicaltrial.gov website, the reference number of the clinical trial is indicated
SMD Stargardt's macular dystrophy, *AMD* age-related macular degeneration, *RP* retinitis pigmentosa

Table 3.3 Clinical trial planned or on-going based on fetal cell types in order to treat RP and/or AMD

Stage of development	Targeted disease	Sponsor/ company	Therapy
Phase I/II NCT02320812	RP	jCyte, Inc.	Human retinal progenitor cells as an intravitreal cell suspension injection
Phase I/II NCT02464436	RP	ReNeuron Limited	Human retinal progenitor cells, as a subretinal cell suspension injection
Phase II NCT02467634	AMD	StemCells, Inc.	Human Central Nervous System Stem Cells (HuCNS-SC), as a subretinal cell suspension injection
Phase I/II NCT01632527	AMD	StemCells, Inc.	Human Central Nervous System Stem Cells (HuCNS-SC), as a subretinal cell suspension injection

When registered in clinicaltrial.gov website, the reference number of the clinical trial is indicated
AMD age-related macular degeneration, *RP* retinitis pigmentosa

reducing their chances of rejection. Thus, the eye appears as a good candidate organ for stem cell therapy. Despite these characteristics, most studies using stem cells for SRS transplantation faced poor survival rate of the graft [168, 169]. The potential for the stem cell-derived RPE cells to replace degenerated endogenous cells in retinal diseases is challenged by this threat of immune rejection. Although immunosuppressive agents were used to address the rejection, significant morbidity is associated with such treatments, especially in the elderly. More knowledge is then necessary about the immune characteristics of the SRS and RPE cells [149].

3.5.1 Immune Privilege of the Eye

The SRS, area between the RPE layer and the outer limiting membrane of the retina, is considered as an immune-privileged site within the eye [170] and thus a logical target for cell transplantation. The integrity of the RPE layer appears to be critical for the immune-privileged status of the SRS [171]. Some studies found that RPE cells in vitro can suppress T-cell activation by direct cell-to-cell contact [172] and by using supernatant of RPE eyecups; others demonstrated that RPE could secrete factors that suppress T-cell activation and production of interferon [173]. RPE possesses characteristics that promote its own survival even when transplanted to nonimmune-privileged site [174]. Indeed CD95-deficient RPE cells (cells lacking Fas Ligand) promote immune reaction leading to the rejection [174]. Transplantation of fetal retina/RPE tissue under the retina of patients suffering from RP or AMD was not associated with significant immune rejection reactions [73]. One explanation proposed by the authors is that the absence of detectable graft rejection, even in patients with donor-specific antibodies before implantation, may indicate that the blood–retinal barrier is restricting antibody access into the SRS. This mechanism could be parallel to the Anterior Chamber-Associated Immune Deviation (ACAID) process existing in the anterior chamber of the eye [73, 163]. Other studies proposed a regulation of T-cell differentiation through TGF-β secretion by RPE cells [175, 176] and/or Interleukin-10 (IL-10) secretion by macrophages [177]. Evidence suggests that innate and adaptive components of the immune system could be regulated through surface expression of molecules on RPE cells, as well as through autocrine and paracrine effects of cytokines and growth factors secreted from the basal and apical sides of the cells [172, 178–181]. The most important molecules secreted by RPE cells and identified to have a role in the regulation of the immune system are (1) TGF-β and thrombospondin (acting on adaptive immune system) and (2) PEDF (Pigment epithelium-derived growth factor) and somatostatin (for innate immune system) [173, 176].

Even if, RPE cells could secrete anti-inflammatory molecules, there are also some evidences that RPE cells could behave as antigen presenting cells to the T cells, stimulating their activation [182]. Regulatory T cells (Tregs) are part of intraocular immunosuppressive mechanisms [183], and their activation depends on TGF-β signaling. To the best of our knowledge, MHC-class II molecules (e.g.,

HLA-DR antigens) are not expressed on RPE cells [181], and there is little or no expression of positive costimulatory molecules (e.g., CD40, CD80, and CD86) by RPE cells under normal conditions [181, 184]. Moreover, RPE cells express negative costimulatory molecules such as B7-H1 (PD-L1), [181] suggesting that T cells infiltrating the graft site after transplantation might interact with these molecules and be inactivated.

Recent studies specifically addressed the capacities of hPSCs-derived RPE cells to regulate subretinal and retinal immune environments. These studies observed that hPSCs-derived RPE cells injected into the SRS survived short-term without evident immune inflammation [116, 185–187]. hiPSCs-derived RPE cell sheets exhibit the morphological properties, in vitro and in vivo function, gene expression, and immunogenicity of authentic RPE cells [116]. In addition, recent clinical trials presumed that hESCs-derived RPE transplants in patients with dry AMD and Stargardt's diseases were not rejected [135, 140]. hiPSCs express low levels of MHC-class I (MHC-I) and β_2-microglobulin (β_2-MG) proteins, which increase upon differentiation [116]. Based on the findings that hiPSCs-derived RPE cells exclusively suppress T-cell activation (e.g., production of IFN-γ) through immunosuppressive factor(s), recent studies showed that cultured hiPSCs-derived RPE cells significantly inhibited cell proliferation and IFN-γ production by T cells when the target T cells were stimulated with anti-human CD3/CD28 antibodies, PHA-P, and recombinant IL-2. The hiPSCs-derived RPE cells constitutively expressed and secreted TGFβ, and TGFβ siRNA-transfected hiPSCs-derived RPE cells did not inhibit T-cell activation. Thus, cultured hiPSCs-derived RPE cells fully suppress T-cell activation in vitro, and hESCs-derived RPE does not stimulate Peripheral Blood Monocytes Cell (PBMC) proliferation (as shown by BrdU assay) nor cytotoxicity (as shown by Mixed Lymphocyte Reaction assay) [116, 188]. However, this is mitigated by a recent study that detected antibody-mediated rejection generated by B cells in allogeneic RPE transplantation in a monkey model [189]. Finally, the maturation status of RPE cells and their organization as polarized cells have effects on the immunosuppressive potential of these cells. Sonoda and collaborators demonstrated that the polarity is associated with a significant shift in growth factors and other cytokines production capacity of the cells [190]. In addition, RPE cells in suspension have a lower tolerance to oxidative stress than those in monolayers [191]. Whether hPSCs-derived RPE cells in suspension culture differ from sheets attached is not clearly determined, but polarized hPSCs-derived RPE might differ in the rate of secretion of molecules involved in immune suppression [116]. Many nonbiodegradable and biodegradable substrates are being tested as scaffold for hPSCs-derived RPE as an alternative to cell suspension. Cons and pros arise from these two strategies as cell suspension confers less surgical trauma but shows a lower tolerance to oxidative stress and a lower survival than epithelial RPE sheets [191].

Even if the SRS is largely an immune-privileged site, attention should be drawn toward the fact that surgical trauma during cell transplantation compromises the blood–ocular barrier and subjects surrounding cells to an increased level of recognition and reactions. Moreover, in the case of diseased patients and particularly in AMD patients where the deregulation of the complement system appears to be a major pathogenesis contributor, promoting the long-term survival of grafted cells in a proinflammatory microenvironment could be a real challenge [149].

In conclusion, the mechanisms conferring immune privilege to the SRS may include (1) suppression of T-cell activation by the release of cytokines from RPE or hPSCs-derived RPE [172, 179, 183, 184, 192–194]; (2) production of other immunosuppressive factors by RPE cells that suppress innate immune activity [173, 194]; (3) surface expression of program death-1 (PD-L1) and Fas ligand [179, 195, 196] by RPE cells; (4) conversion of CD8+ and CD4+ T cells into regulatory T cells [193]; and (5) the intact physical barrier of the RPE layer [171].

3.5.2 Current Immunosuppressive Treatment and Global Haplotype Cell Banking Initiative

Occurrence of chronic loss of transplanted RPE cells in the SRS indicates that the immune privilege is not perfect despite these immunosuppressive mechanisms [163, 197]. Several laboratories demonstrated that even with robust anti-inflammatory regimen, gradual loss of allogeneic or xenografted RPE could occur and starts 1–2 months after grafting into the SRS [83, 198, 199]. Local inflammation can be induced by numerous factors among which any artificial scaffold associated to the transplanted cells.

In order to obtain long-term survival and function of the transplanted cells, we might provide some protection against the inflammatory response that could be triggered at the time of surgery by using robust immunosuppressive regimen (mycophenolate mofetil and tacrolimus, [135, 140]) or intraocular corticosteroid capsules (intravitreal dexamethasone or fluocinolone acetonide implants [200, 201]).

Matching MHC, ABO blood groups and minor antigens improve the survival of hiPSCs-derived RPE grafts [116]. However, allofactors other than MHC antigens of the transplanted cells are recognized as a heterogeneous group of targets for immune recognition and the importance of such molecules is illustrated by the observation of transplant rejection in HLA identical siblings [202].

Generation of hPSCs-derived RPE with known HLA genotypes may reduce risks of rejection [141, 203]. This concept already exists for cord blood [204]. HLA genes are located on chromosome 6 and represent the most polymorphic system in the human genome. They are divided into two groups of antigens, HLA class I or HLA-I (HLA-A, HLA-B, and HLA-C) expressed by nearly all nucleated cells and HLA class II or HLA-II (HLA-DR, HLA-DP, and HLA-DQ), which are expressed by some specific cells, such as dendritic or B cells [205]. Studies are now conducted to determine whether allogeneic T cells can recognize hiPSCs-derived RPE cells from HLA-3 locus homozygote donors [206]. In solid organ transplantation, many studies have demonstrated the importance of HLA-A, HLA-B, and HLA-DR for the long-term graft survival in addition to immunosuppressive drugs. Matching for these loci not only reduces the allograft rejection but also diminishes the use of immunosuppressive drugs [207]. Sugita and collaborators demonstrated that hiPSCs-derived RPE from HLA homozygous did not elicit a T-cell response in vitro [208]. This group also demonstrated that MHC homozygous donor of iPSCs-derived RPE transplanted to a compatible recipient monkey did not induced the infiltration of inflammatory cells to the graft. By contrast, MHC mismatched donor and recipient monkeys had inflammatory cell infiltration into the transplanted RPE sheet [209]. Due to HLA diversity, the best possibility is a bank of HLA homozygous cell lines. Nakajima et al. calculated that if a bank possesses hPSC lines from 100 randomly healthy selected donated embryo, 19% of patients were expected to find at least one hiPS cell line with a complete matching for HLA-A, HLA-B, and HLA-DR in the Japanese population [210]. Taylor and collaborators showed that, in the UK, 150 donors would cover 18.5% of the population with a full match. Interestingly, using this number of 150 donors would also cover 21% of the Japanese population with a full match [203]. Gourraud and collaborators developed a probabilistic model and demonstrated that using a bank comprising the 100 iPSC lines with the most frequent HLA in each population would leave out only 22% of the European Americans, but 37% of the Asians, 48% of the Hispanics, and 55% of the African Americans [211]. International strategies started now in order to create such banks, which might be useful as a source of allografts in retinal disorders [141].

3.6 Conclusions

Human retinal degenerative diseases are currently incurable, and retinal degeneration, once initiated, is irreversible. As RPE cell defects are involved in some RP and AMD pathogenesis, reestablishment of a healthy RPE layer through implantation of hPSCs-derived RPE has created hope for preventing blindness of millions of patients. Recent US clinical trials demonstrated the safety of nonpolarized hESCs-derived RPE approaches, and other ongoing clinical trials are promising in evaluating functional outcomes for nonpolarized and polarized hPSCs-derived RPE transplantation. As pointed out by the NEI, technical and logistical roadblocks remain to be identified, and potential solutions will come from collaboration between academic labs and private companies [157]. We are now entering a very exciting era leading to different treatment options for patients.

References

1. Wong WL, Su X, Li X, Cheung CM, Klein R, Cheng CY, Wong TY (2014) Global prevalence of age-related macular degeneration and disease burden projection for 2020 and 2040: a systematic review and meta-analysis. Lancet Glob Health 2(2):e106–e116
2. Kolb H, Nelson R, Fernandez E, Jones B (2012) WEBVISION, The Organization of the Retina and Visual System. http://webvision.med.utah.edu/

3. Nag TC, Wadhwa S (2012) Ultrastructure of the human retina in aging and various pathological states. Micron 43(7):759–781
4. Sparrow JR, Hicks D, Hamel CP (2010) The retinal pigment epithelium in health and disease. Curr Mol Med 10(9):802–823
5. Cuenca N, Fernandez-Sanchez L, Campello L, Maneu V, De la Villa P, Lax P, Pinilla I (2014) Cellular responses following retinal injuries and therapeutic approaches for neurodegenerative diseases. Prog Retin Eye Res 43:17–75
6. Strauss O (2005) The retinal pigment epithelium in visual function. Physiol Rev 85(3):845–881
7. Ben M'Barek K, Regent F, Monville C (2015) Use of human pluripotent stem cells to study and treat retinopathies. World J Stem Cells 7(3):596–604
8. da Cruz L, Chen FK, Ahmado A, Greenwood J, Coffey P (2007) RPE transplantation and its role in retinal disease. Prog Retin Eye Res 26(6):598–635
9. Bird AC (1995) Retinal photoreceptor dystrophies LI. Edward Jackson Memorial Lecture. Am J Ophthalmol 119(5):543–562
10. Sahel JA, Marazova K, Audo I (2015) Clinical characteristics and current therapies for inherited retinal degenerations. Cold Spring Harb Perspect Med 5(2):a017111
11. Hartong DT, Berson EL, Dryja TP (2006) Retinitis pigmentosa. Lancet 368(9549):1795–1809
12. Berger W, Kloeckener-Gruissem B, Neidhardt J (2010) The molecular basis of human retinal and vitreoretinal diseases. Prog Retin Eye Res 29(5):335–375
13. den Hollander AI, Roepman R, Koenekoop RK, Cremers FP (2008) Leber congenital amaurosis: genes, proteins and disease mechanisms. Prog Retin Eye Res 27(4):391–419
14. Hamel CP (2014) Gene discovery and prevalence in inherited retinal dystrophies. C R Biol 337(3):160–166
15. Hamel CP, Griffoin JM, Lasquellec L, Bazalgette C, Arnaud B (2001) Retinal dystrophies caused by mutations in RPE65: assessment of visual functions. Br J Ophthalmol 85(4):424–427
16. Marlhens F, Bareil C, Griffoin JM, Zrenner E, Amalric P, Eliaou C, Liu SY, Harris E, Redmond TM, Arnaud B, Claustres M, Hamel CP (1997) Mutations in RPE65 cause Leber's congenital amaurosis. Nat Genet 17(2):139–141
17. Marlhens F, Griffoin JM, Bareil C, Arnaud B, Claustres M, Hamel CP (1998) Autosomal recessive retinal dystrophy associated with two novel mutations in the RPE65 gene. Eur J Hum Genet 6(5):527–531
18. Morimura H, Fishman GA, Grover SA, Fulton AB, Berson EL, Dryja TP (1998) Mutations in the RPE65 gene in patients with autosomal recessive retinitis pigmentosa or leber congenital amaurosis. Proc Natl Acad Sci U S A 95(6):3088–3093
19. Burstedt MS, Forsman-Semb K, Golovleva I, Janunger T, Wachtmeister L, Sandgren O (2001) Ocular phenotype of bothnia dystrophy, an autosomal recessive retinitis pigmentosa associated with an R234W mutation in the RLBP1 gene. Arch Ophthalmol 119(2):260–267
20. Maw MA, Kennedy B, Knight A, Bridges R, Roth KE, Mani EJ, Mukkadan JK, Nancarrow D, Crabb JW, Denton MJ (1997) Mutation of the gene encoding cellular retinaldehyde-binding protein in autosomal recessive retinitis pigmentosa. Nat Genet 17(2):198–200
21. Young RW, Bok D (1969) Participation of the retinal pigment epithelium in the rod outer segment renewal process. J Cell Biol 42(2):392–403
22. Bok D (1985) Retinal photoreceptor-pigment epithelium interactions. Friedenwald lecture. Invest Ophthalmol Vis Sci 26(12):1659–1694
23. Tarnowski BI, Shepherd VL, McLaughlin BJ (1988) Mannose 6-phosphate receptors on the plasma membrane on rat retinal pigment epithelial cells. Invest Ophthalmol Vis Sci 29(2):291–297
24. Ryeom SW, Silverstein RL, Scotto A, Sparrow JR (1996) Binding of anionic phospholipids to retinal pigment epithelium may be mediated by the scavenger receptor CD36. J Biol Chem 271(34):20536–20539
25. Ryeom SW, Sparrow JR, Silverstein RL (1996) CD36 participates in the phagocytosis of rod outer segments by retinal pigment epithelium. J Cell Sci 109(Pt 2):387–395

26. Lin H, Clegg DO (1998) Integrin alphavbeta5 participates in the binding of photoreceptor rod outer segments during phagocytosis by cultured human retinal pigment epithelium. Invest Ophthalmol Vis Sci 39(9):1703–1712
27. Hall MO, Burgess BL, Abrams TA, Ershov AV, Gregory CY (1996) Further studies on the identification of the phagocytosis receptor of rat retinal pigment epithelial cells. Exp Eye Res 63(3):255–264
28. D'Cruz PM, Yasumura D, Weir J, Matthes MT, Abderrahim H, LaVail MM, Vollrath D (2000) Mutation of the receptor tyrosine kinase gene Mertk in the retinal dystrophic RCS rat. Hum Mol Genet 9(4):645–651
29. Nandrot EF, Kim Y, Brodie SE, Huang X, Sheppard D, Finnemann SC (2004) Loss of synchronized retinal phagocytosis and age-related blindness in mice lacking alphavbeta5 integrin. J Exp Med 200(12):1539–1545
30. Gal A, Li Y, Thompson DA, Weir J, Orth U, Jacobson SG, Apfelstedt-Sylla E, Vollrath D (2000) Mutations in MERTK, the human orthologue of the RCS rat retinal dystrophy gene, cause retinitis pigmentosa. Nat Genet 26(3):270–271
31. Mazzoni F, Safa H, Finnemann SC (2014) Understanding photoreceptor outer segment phagocytosis: use and utility of RPE cells in culture. Exp Eye Res 126:51–60
32. Zarbin MA, Rosenfeld PJ (2010) Pathway-based therapies for age-related macular degeneration: an integrated survey of emerging treatment alternatives. Retina 30(9):1350–1367
33. Ambati J, Ambati BK, Yoo SH, Ianchulev S, Adamis AP (2003) Age-related macular degeneration: etiology, pathogenesis, and therapeutic strategies. Surv Ophthalmol 48(3):257–293
34. Boulton M, Dayhaw-Barker P (2001) The role of the retinal pigment epithelium: topographical variation and ageing changes. Eye (Lond) 15(Pt 3):384–389
35. Nowak JZ (2006) Age-related macular degeneration (AMD): pathogenesis and therapy. Pharmacol Rep 58(3):353–363
36. Xu H, Chen M, Forrester JV (2009) Para-inflammation in the aging retina. Prog Retin Eye Res 28(5):348–368
37. Bhutto I, Lutty G (2012) Understanding age-related macular degeneration (AMD): relationships between the photoreceptor/retinal pigment epithelium/Bruch's membrane/choriocapillaris complex. Mol Aspects Med 33(4):295–317
38. Dunaief JL, Dentchev T, Ying GS, Milam AH (2002) The role of apoptosis in age-related macular degeneration. Arch Ophthalmol 120(11):1435–1442
39. Fritsche LG, Chen W, Schu M, Yaspan BL, Yu Y, Thorleifsson G, Zack DJ, Arakawa S, Cipriani V, Ripke S, Igo RP Jr, Buitendijk GH, Sim X, Weeks DE, Guymer RH, Merriam JE, Francis PJ, Hannum G, Agarwal A, Armbrecht AM, Audo I, Aung T, Barile GR, Benchaboune M, Bird AC, Bishop PN, Branham KE, Brooks M, Brucker AJ, Cade WH, Cain MS, Campochiaro PA, Chan CC, Cheng CY, Chew EY, Chin KA, Chowers I, Clayton DG, Cojocaru R, Conley YP, Cornes BK, Daly MJ, Dhillon B, Edwards AO, Evangelou E, Fagerness J, Ferreyra HA, Friedman JS, Geirsdottir A, George RJ, Gieger C, Gupta N, Hagstrom SA, Harding SP, Haritoglou C, Heckenlively JR, Holz FG, Hughes G, Ioannidis JP, Ishibashi T, Joseph P, Jun G, Kamatani Y, Katsanis N, Keilhauer N, Khan JC, Kim IK, Kiyohara Y, Klein BE, Klein R, Kovach JL, Kozak I, Lee CJ, Lee KE, Lichtner P, Lotery AJ, Meitinger T, Mitchell P, Mohand-Said S, Moore AT, Morgan DJ, Morrison MA, Myers CE, Naj AC, Nakamura Y, Okada Y, Orlin A, Ortube MC, Othman MI, Pappas C, Park KH, Pauer GJ, Peachey NS, Poch O, Priya RR, Reynolds R, Richardson AJ, Ripp R, Rudolph G, Ryu E, Sahel JA, Schaumberg DA, Scholl HP, Schwartz SG, Scott WK, Shahid H, Sigurdsson H, Silvestri G, Sivakumaran TA, Smith RT, Sobrin L, Souied EH, Stambolian DE, Stefansson H, Sturgill-Short GM, Takahashi A, Tosakulwong N, Truitt BJ, Tsironi EE, Uitterlinden AG, van Duijn CM, Vijaya L, Vingerling JR, Vithana EN, Webster AR, Wichmann HE, Winkler TW, Wong TY, Wright AF, Zelenika D, Zhang M, Zhao L, Zhang K, Klein ML, Hageman GS, Lathrop GM, Stefansson K, Allikmets R, Baird PN, Gorin MB, Wang JJ, Klaver CC, Seddon JM, Pericak-Vance MA, Iyengar SK, Yates JR, Swaroop A, Weber BH, Kubo M, Deangelis MM, Leveillard T, Thorsteinsdottir U, Haines JL, Farrer LA, Heid IM, Abecasis GR, A. M.

D. G. Consortium (2013) Seven new loci associated with age-related macular degeneration. Nat Genet 45(4):433–439, 439e431–432

40. Macular Photocoagulation Study Group (1991) Laser photocoagulation of subfoveal neovascular lesions in age-related macular degeneration. Results of a randomized clinical trial. Macular Photocoagulation Study Group. Arch Ophthalmol 109(9):1220–1231

41. Azab M, Boyer DS, Bressler NM, Bressler SB, Cihelkova I, Hao Y, Immonen I, Lim JI, Menchini U, Naor J, Potter MJ, Reaves A, Rosenfeld PJ, Slakter JS, Soucek P, Strong HA, Wenkstern A, Su XY, Yang YC, Visudyne in Minimally Classic Choroidal Neovascularization Study Group (2005) Verteporfin therapy of subfoveal minimally classic choroidal neovascularization in age-related macular degeneration: 2-year results of a randomized clinical trial. Arch Ophthalmol 123(4):448–457

42. Gragoudas ES, Adamis AP, Cunningham ET Jr, Feinsod M, Guyer DR, VEGF Inhibition Study in Ocular Neovascularization Clinical Trial (2004) Pegaptanib for neovascular age-related macular degeneration. N Engl J Med 351(27):2805–2816

43. Brown DM, Michels M, Kaiser PK, Heier JS, Sy JP, Ianchulev T, A. S. Group (2009) Ranibizumab versus verteporfin photodynamic therapy for neovascular age-related macular degeneration: two-year results of the ANCHOR study. Ophthalmology 116(1):57–65.e55

44. Chang TS, Bressler NM, Fine JT, Dolan CM, Ward J, Klesert TR, M. S. Group (2007) Improved vision-related function after ranibizumab treatment of neovascular age-related macular degeneration: results of a randomized clinical trial. Arch Ophthalmol 125(11):1460–1469

45. Rosenfeld PJ, Brown DM, Heier JS, Boyer DS, Kaiser PK, Chung CY, Kim RY, M. S. Group (2006) Ranibizumab for neovascular age-related macular degeneration. N Engl J Med 355(14):1419–1431

46. Schmidt-Erfurth U, Kaiser PK, Korobelnik JF, Brown DM, Chong V, Nguyen QD, Ho AC, Ogura Y, Simader C, Jaffe GJ, Slakter JS, Yancopoulos GD, Stahl N, Vitti R, Berliner AJ, Soo Y, Anderesi M, Sowade O, Zeitz O, Norenberg C, Sandbrink R, Heier JS (2014) Intravitreal aflibercept injection for neovascular age-related macular degeneration: ninety-six-week results of the VIEW studies. Ophthalmology 121(1):193–201

47. CATT Research Group, Martin DF, Maguire MG, Ying GS, Grunwald JE, Fine SL, Jaffe GJ (2011) Ranibizumab and bevacizumab for neovascular age-related macular degeneration. N Engl J Med 364(20):1897–1908

48. Rosenfeld PJ (2006) Intravitreal avastin: the low cost alternative to lucentis? Am J Ophthalmol 142(1):141–143

49. Steinbrook R (2006) The price of sight—ranibizumab, bevacizumab, and the treatment of macular degeneration. N Engl J Med 355(14):1409–1412

50. Tolentino MJ, Dennrick A, John E, Tolentino MS (2015) Drugs in Phase II clinical trials for the treatment of age-related macular degeneration. Expert Opin Investig Drugs 24(2):183–199

51. Cideciyan AV, Jacobson SG, Beltran WA, Sumaroka A, Swider M, Iwabe S, Roman AJ, Olivares MB, Schwartz SB, Komaromy AM, Hauswirth WW, Aguirre GD (2013) Human retinal gene therapy for Leber congenital amaurosis shows advancing retinal degeneration despite enduring visual improvement. Proc Natl Acad Sci U S A 110(6):E517–E525

52. Ali RR (2012) Gene therapy for retinal dystrophies: twenty years in the making. Hum Gene Ther 23(4):337–339

53. Bainbridge JW, Mehat MS, Sundaram V, Robbie SJ, Barker SE, Ripamonti C, Georgiadis A, Mowat FM, Beattie SG, Gardner PJ, Feathers KL, Luong VA, Yzer S, Balaggan K, Viswanathan A, de Ravel TJ, Casteels I, Holder GE, Tyler N, Fitzke FW, Weleber RG, Nardini M, Moore AT, Thompson DA, Petersen-Jones SM, Michaelides M, van den Born LI, Stockman A, Smith AJ, Rubin G, Ali RR (2015) Long-term effect of gene therapy on Leber's congenital amaurosis. N Engl J Med 372(20):1887–1897

54. Jha BS, Bharti K (2015) Regenerating retinal pigment epithelial cells to cure blindness: a road towards personalized artificial tissue. Curr Stem Cell Rep 1(2):79–91

55. Gouras P, Flood MT, Kjedbye H, Bilek MK, Eggers H (1985) Transplantation of cultured human retinal epithelium to Bruch's membrane of the owl monkey's eye. Curr Eye Res 4(3):253–265

56. Gouras P, Flood MT, Kjeldbye H (1984) Transplantation of cultured human retinal cells to monkey retina. An Acad Bras Cienc 56(4):431–443

57. Lopez R, Gouras P, Kjeldbye H, Sullivan B, Reppucci V, Brittis M, Wapner F, Goluboff E (1989) Transplanted retinal pigment epithelium modifies the retinal degeneration in the RCS rat. Invest Ophthalmol Vis Sci 30(3):586–588

58. World Medical Association (2013) World Medical Association Declaration of Helsinki: ethical principles for medical research involving human subjects. JAMA 310(20):2191–2194

59. Liao JL, Yu J, Huang K, Hu J, Diemer T, Ma Z, Dvash T, Yang XJ, Travis GH, Williams DS, Bok D, Fan G (2010) Molecular signature of primary retinal pigment epithelium and stem-cell-derived RPE cells. Hum Mol Genet 19(21):4229–4238

60. Maminishkis A, Chen S, Jalickee S, Banzon T, Shi G, Wang FE, Ehalt T, Hammer JA, Miller SS (2006) Confluent monolayers of cultured human fetal retinal pigment epithelium exhibit morphology and physiology of native tissue. Invest Ophthalmol Vis Sci 47(8):3612–3624

61. Zhu M, Provis JM, Penfold PL (1998) Isolation, culture and characteristics of human foetal and adult retinal pigment epithelium. Aust N Z J Ophthalmol 26(Suppl 1):S50–S52

62. Nandrot EF, Dufour EM (2010) Mertk in daily retinal phagocytosis: a history in the making. Adv Exp Med Biol 664:133–140

63. Mullen RJ, LaVail MM (1976) Inherited retinal dystrophy: primary defect in pigment epithelium determined with experimental rat chimeras. Science 192(4241):799–801

64. Little CW, Castillo B, DiLoreto DA, Cox C, Wyatt J, del Cerro C, del Cerro M (1996) Transplantation of human fetal retinal pigment epithelium rescues photoreceptor cells from degeneration in the Royal College of Surgeons rat retina. Invest Ophthalmol Vis Sci 37(1):204–211

65. Sheng Y, Gouras P, Cao H, Berglin L, Kjeldbye H, Lopez R, Rosskothen H (1995) Patch transplants of human fetal retinal pigment epithelium in rabbit and monkey retina. Invest Ophthalmol Vis Sci 36(2):381–390

66. Berglin L, Gouras P, Sheng Y, Lavid J, Lin PK, Cao H, Kjeldbye H (1997) Tolerance of human fetal retinal pigment epithelium xenografts in monkey retina. Graefes Arch Clin Exp Ophthalmol 235(2):103–110

67. Lai CC, Gouras P, Doi K, Tsang SH, Goff SP, Ashton P (2000) Local immunosuppression prolongs survival of RPE xenografts labeled by retroviral gene transfer. Invest Ophthalmol Vis Sci 41(10):3134–3141

68. Aramant RB, Seiler MJ (2002) Transplanted sheets of human retina and retinal pigment epithelium develop normally in nude rats. Exp Eye Res 75(2):115–125

69. Algvere PV, Berglin L, Gouras P, Sheng Y (1994) Transplantation of fetal retinal pigment epithelium in age-related macular degeneration with subfoveal neovascularization. Graefes Arch Clin Exp Ophthalmol 232(12):707–716

70. Algvere PV, Gouras P, Dafgard Kopp E (1999) Long-term outcome of RPE allografts in non-immunosuppressed patients with AMD. Eur J Ophthalmol 9(3):217–230

71. Weisz JM, Humayun MS, De Juan E Jr, Del Cerro M, Sunness JS, Dagnelie G, Soylu M, Rizzo L, Nussenblatt RB (1999) Allogenic fetal retinal pigment epithelial cell transplant in a patient with geographic atrophy. Retina 19(6):540–545

72. Radtke ND, Seiler MJ, Aramant RB, Petry HM, Pidwell DJ (2002) Transplantation of intact sheets of fetal neural retina with its retinal pigment epithelium in retinitis pigmentosa patients. Am J Ophthalmol 133(4):544–550

73. Radtke ND, Aramant RB, Petry HM, Green PT, Pidwell DJ, Seiler MJ (2008) Vision improvement in retinal degeneration patients by implantation of retina together with retinal pigment epithelium. Am J Ophthalmol 146(2):172–182

74. Binder S, Stanzel BV, Krebs I, Glittenberg C (2007) Transplantation of the RPE in AMD. Prog Retin Eye Res 26(5):516–554

75. Castillo BV Jr, del Cerro M, White RM, Cox C, Wyatt J, Nadiga G, del Cerro C (1997) Efficacy of nonfetal human RPE for photoreceptor rescue: a study in dystrophic RCS rats. Exp Neurol 146(1):1–9
76. Phillips SJ, Sadda SR, Tso MO, Humayan MS, de Juan E Jr, Binder S (2003) Autologous transplantation of retinal pigment epithelium after mechanical debridement of Bruch's membrane. Curr Eye Res 26(2):81–88
77. Binder S, Stolba U, Krebs I, Kellner L, Jahn C, Feichtinger H, Povelka M, Frohner U, Kruger A, Hilgers RD, Krugluger W (2002) Transplantation of autologous retinal pigment epithelium in eyes with foveal neovascularization resulting from age-related macular degeneration: a pilot study. Am J Ophthalmol 133(2):215–225
78. Dunn KC, Aotaki-Keen AE, Putkey FR, Hjelmeland LM (1996) ARPE-19, a human retinal pigment epithelial cell line with differentiated properties. Exp Eye Res 62(2):155–169
79. Pinilla I, Cuenca N, Sauve Y, Wang S, Lund RD (2007) Preservation of outer retina and its synaptic connectivity following subretinal injections of human RPE cells in the Royal College of Surgeons rat. Exp Eye Res 85(3):381–392
80. Coffey PJ, Girman S, Wang SM, Hetherington L, Keegan DJ, Adamson P, Greenwood J, Lund RD (2002) Long-term preservation of cortically dependent visual function in RCS rats by transplantation. Nat Neurosci 5(1):53–56
81. Salero E, Blenkinsop TA, Corneo B, Harris A, Rabin D, Stern JH, Temple S (2012) Adult human RPE can be activated into a multipotent stem cell that produces mesenchymal derivatives. Cell Stem Cell 10(1):88–95
82. Chiba C (2014) The retinal pigment epithelium: an important player of retinal disorders and regeneration. Exp Eye Res 123:107–114
83. Stanzel BV, Liu Z, Somboonthanakij S, Wongsawad W, Brinken R, Eter N, Corneo B, Holz FG, Temple S, Stern JH, Blenkinsop TA (2014) Human RPE stem cells grown into polarized RPE monolayers on a polyester matrix are maintained after grafting into rabbit subretinal space. Stem Cell Rep 2(1):64–77
84. Schraermeyer U, Enzmann V, Kohen L, Addicks K, Wiedemann P, Heimann K (1997) Porcine iris pigment epithelial cells can take up retinal outer segments. Exp Eye Res 65(2):277–287
85. Schraermeyer U, Kociok N, Heimann K (1999) Rescue effects of IPE transplants in RCS rats: short-term results. Invest Ophthalmol Vis Sci 40(7):1545–1556
86. Thumann G, Salz AK, Walter P, Johnen S (2009) Preservation of photoreceptors in dystrophic RCS rats following allo- and xenotransplantation of IPE cells. Graefes Arch Clin Exp Ophthalmol 247(3):363–369
87. Abe T, Yoshida M, Tomita H, Kano T, Sato M, Wada Y, Fuse N, Yamada T, Tamai M (2000) Auto iris pigment epithelial cell transplantation in patients with age-related macular degeneration: short-term results. Tohoku J Exp Med 191(1):7–20
88. Abe T, Yoshida M, Yoshioka Y, Wakusawa R, Tokita-Ishikawa Y, Seto H, Tamai M, Nishida K (2007) Iris pigment epithelial cell transplantation for degenerative retinal diseases. Prog Retin Eye Res 26(3):302–321
89. Semkova I, Kreppel F, Welsandt G, Luther T, Kozlowski J, Janicki H, Kochanek S, Schraermeyer U (2002) Autologous transplantation of genetically modified iris pigment epithelial cells: a promising concept for the treatment of age-related macular degeneration and other disorders of the eye. Proc Natl Acad Sci U S A 99(20):13090–13095
90. Hojo M, Abe T, Sugano E, Yoshioka Y, Saigo Y, Tomita H, Wakusawa R, Tamai M (2004) Photoreceptor protection by iris pigment epithelial transplantation transduced with AAV-mediated brain-derived neurotrophic factor gene. Invest Ophthalmol Vis Sci 45(10):3721–3726
91. Lawrence JM, Sauve Y, Keegan DJ, Coffey PJ, Hetherington L, Girman S, Whiteley SJ, Kwan AS, Pheby T, Lund RD (2000) Schwann cell grafting into the retina of the dystrophic RCS rat limits functional deterioration. Royal College of Surgeons. Invest Ophthalmol Vis Sci 41(2):518–528
92. Keegan DJ, Kenna P, Humphries MM, Humphries P, Flitcroft DI, Coffey PJ, Lund RD, Lawrence JM (2003) Transplantation of syngeneic Schwann cells to the retina of the rhodopsin knockout (rho(−/−)) mouse. Invest Ophthalmol Vis Sci 44(8):3526–3532

93. Lawrence JM, Keegan DJ, Muir EM, Coffey PJ, Rogers JH, Wilby MJ, Fawcett JW, Lund RD (2004) Transplantation of Schwann cell line clones secreting GDNF or BDNF into the retinas of dystrophic Royal College of Surgeons rats. Invest Ophthalmol Vis Sci 45(1):267–274

94. Gamm DM, Wang S, Lu B, Girman S, Holmes T, Bischoff N, Shearer RL, Sauve Y, Capowski E, Svendsen CN, Lund RD (2007) Protection of visual functions by human neural progenitors in a rat model of retinal disease. PLoS One 2(3):e338

95. Wang S, Girman S, Lu B, Bischoff N, Holmes T, Shearer R, Wright LS, Svendsen CN, Gamm DM, Lund RD (2008) Long-term vision rescue by human neural progenitors in a rat model of photoreceptor degeneration. Invest Ophthalmol Vis Sci 49(7):3201–3206

96. Francis PJ, Wang S, Zhang Y, Brown A, Hwang T, McFarland TJ, Jeffrey BG, Lu B, Wright L, Appukuttan B, Wilson DJ, Stout JT, Neuringer M, Gamm DM, Lund RD (2009) Subretinal transplantation of forebrain progenitor cells in nonhuman primates: survival and intact retinal function. Invest Ophthalmol Vis Sci 50(7):3425–3431

97. Tsukamoto A, Uchida N, Capela A, Gorba T, Huhn S (2013) Clinical translation of human neural stem cells. Stem Cell Res Ther 4(4):102

98. McGill TJ, Cottam B, Lu B, Wang S, Girman S, Tian C, Huhn SL, Lund RD, Capela A (2012) Transplantation of human central nervous system stem cells - neuroprotection in retinal degeneration. Eur J Neurosci 35(3):468–477

99. Casarosa S, Bozzi Y, Conti L (2014) Neural stem cells: ready for therapeutic applications? Mol Cell Ther 2:31

100. Chen Y, Shao JZ, Xiang LX, Dong XJ, Zhang GR (2008) Mesenchymal stem cells: a promising candidate in regenerative medicine. Int J Biochem Cell Biol 40(5):815–820

101. Chen Q, Long Y, Yuan X, Zou L, Sun J, Chen S, Perez-Polo JR, Yang K (2005) Protective effects of bone marrow stromal cell transplantation in injured rodent brain: synthesis of neurotrophic factors. J Neurosci Res 80(5):611–619

102. Arnhold S, Heiduschka P, Klein H, Absenger Y, Basnaoglu S, Kreppel F, Henke-Fahle S, Kochanek S, Bartz-Schmidt KU, Addicks K, Schraermeyer U (2006) Adenovirally transduced bone marrow stromal cells differentiate into pigment epithelial cells and induce rescue effects in RCS rats. Invest Ophthalmol Vis Sci 47(9):4121–4129

103. Inoue Y, Iriyama A, Ueno S, Takahashi H, Kondo M, Tamaki Y, Araie M, Yanagi Y (2007) Subretinal transplantation of bone marrow mesenchymal stem cells delays retinal degeneration in the RCS rat model of retinal degeneration. Exp Eye Res 85(2):234–241

104. Lu B, Wang S, Girman S, McGill T, Ragaglia V, Lund R (2010) Human adult bone marrow-derived somatic cells rescue vision in a rodent model of retinal degeneration. Exp Eye Res 91(3):449–455

105. Tzameret A, Sher I, Belkin M, Treves AJ, Meir A, Nagler A, Levkovitch-Verbin H, Barshack I, Rosner M, Rotenstreich Y (2014) Transplantation of human bone marrow mesenchymal stem cells as a thin subretinal layer ameliorates retinal degeneration in a rat model of retinal dystrophy. Exp Eye Res 118:135–144

106. Ahmad I, Tang L, Pham H (2000) Identification of neural progenitors in the adult mammalian eye. Biochem Biophys Res Commun 270(2):517–521

107. Ballios BG, Clarke L, Coles BL, Shoichet MS, Van Der Kooy D (2012) The adult retinal stem cell is a rare cell in the ciliary epithelium whose progeny can differentiate into photoreceptors. Biol Open 1(3):237–246

108. Tropepe V, Coles BL, Chiasson BJ, Horsford DJ, Elia AJ, McInnes RR, van der Kooy D (2000) Retinal stem cells in the adult mammalian eye. Science 287(5460):2032–2036

109. Coles BL, Angenieux B, Inoue T, Del Rio-Tsonis K, Spence JR, McInnes RR, Arsenijevic Y, van der Kooy D (2004) Facile isolation and the characterization of human retinal stem cells. Proc Natl Acad Sci U S A 101(44):15772–15777

110. Cicero SA, Johnson D, Reyntjens S, Frase S, Connell S, Chow LM, Baker SJ, Sorrentino BP, Dyer MA (2009) Cells previously identified as retinal stem cells are pigmented ciliary epithelial cells. Proc Natl Acad Sci U S A 106(16):6685–6690

111. Froen R, Johnsen EO, Nicolaissen B, Facsko A, Petrovski G, Moe MC (2013) Does the adult human ciliary body epithelium contain "true" retinal stem cells? Biomed Res Int 2013:531579

112. Lund RD, Wang S, Lu B, Girman S, Holmes T, Sauve Y, Messina DJ, Harris IR, Kihm AJ, Harmon AM, Chin FY, Gosiewska A, Mistry SK (2007) Cells isolated from umbilical cord tissue rescue photoreceptors and visual functions in a rodent model of retinal disease. Stem Cells 25(3):602–611

113. Thomson JA, Itskovitz-Eldor J, Shapiro SS, Waknitz MA, Swiergiel JJ, Marshall VS, Jones JM (1998) Embryonic stem cell lines derived from human blastocysts. Science 282(5391):1145–1147

114. Takahashi K, Tanabe K, Ohnuki M, Narita M, Ichisaka T, Tomoda K, Yamanaka S (2007) Induction of pluripotent stem cells from adult human fibroblasts by defined factors. Cell 131(5):861–872

115. Yu J, Vodyanik MA, Smuga-Otto K, Antosiewicz-Bourget J, Frane JL, Tian S, Nie J, Jonsdottir GA, Ruotti V, Stewart R, Slukvin II, Thomson JA (2007) Induced pluripotent stem cell lines derived from human somatic cells. Science 318(5858):1917–1920

116. Kamao H, Mandai M, Okamoto S, Sakai N, Suga A, Sugita S, Kiryu J, Takahashi M (2014) Characterization of human induced pluripotent stem cell-derived retinal pigment epithelium cell sheets aiming for clinical application. Stem Cell Rep 2(2):205–218

117. Leach LL, Clegg DO (2015) Concise review: Making stem cells retinal: methods for deriving retinal pigment epithelium and implications for patients with ocular disease. Stem Cells 33(8):2363–2373

118. Klimanskaya I, Hipp J, Rezai KA, West M, Atala A, Lanza R (2004) Derivation and comparative assessment of retinal pigment epithelium from human embryonic stem cells using transcriptomics. Cloning Stem Cells 6(3):217–245

119. Choudhary P, Booth H, Gutteridge A, Surmacz B, Louca I, Steer J, Kerby J, Whiting PJ (2017) Directing differentiation of pluripotent stem cells toward retinal pigment epithelium lineage. Stem Cells Transl Med 6(2):490–501

120. Lidgerwood GE, Lim SY, Crombie DE, Ali R, Gill KP, Hernandez D, Kie J, Conquest A, Waugh HS, Wong RC, Liang HH, Hewitt AW, Davidson KC, Pebay A (2016) Defined medium conditions for the induction and expansion of human pluripotent stem cell-derived retinal pigment epithelium. Stem Cell Rev 12(2):179–188

121. Lamba DA, Karl MO, Ware CB, Reh TA (2006) Efficient generation of retinal progenitor cells from human embryonic stem cells. Proc Natl Acad Sci U S A 103(34):12769–12774

122. Buchholz DE, Pennington BO, Croze RH, Hinman CR, Coffey PJ, Clegg DO (2013) Rapid and efficient directed differentiation of human pluripotent stem cells into retinal pigmented epithelium. Stem Cells Transl Med 2(5):384–393

123. Leach LL, Buchholz DE, Nadar VP, Lowenstein SE, Clegg DO (2015) Canonical/beta-catenin Wnt pathway activation improves retinal pigmented epithelium derivation from human embryonic stem cells. Invest Ophthalmol Vis Sci 56(2):1002–1013

124. Bharti K, Miller SS, Arnheiter H (2011) The new paradigm: retinal pigment epithelium cells generated from embryonic or induced pluripotent stem cells. Pigment Cell Melanoma Res 24(1):21–34

125. Buchholz DE, Hikita ST, Rowland TJ, Friedrich AM, Hinman CR, Johnson LV, Clegg DO (2009) Derivation of functional retinal pigmented epithelium from induced pluripotent stem cells. Stem Cells 27(10):2427–2434

126. Maruotti J, Sripathi SR, Bharti K, Fuller J, Wahlin KJ, Ranganathan V, Sluch VM, Berlinicke CA, Davis J, Kim C, Zhao L, Wan J, Qian J, Corneo B, Temple S, Dubey R, Olenyuk BZ, Bhutto I, Lutty GA, Zack DJ (2015) Small-molecule-directed, efficient generation of retinal pigment epithelium from human pluripotent stem cells. Proc Natl Acad Sci U S A 112(35):10950–10955

127. Odorico JS, Kaufman DS, Thomson JA (2001) Multilineage differentiation from human embryonic stem cell lines. Stem Cells 19(3):193–204

128. Itskovitz-Eldor J, Schuldiner M, Karsenti D, Eden A, Yanuka O, Amit M, Soreq H, Benvenisty N (2000) Differentiation of human embryonic stem cells into embryoid bodies compromising the three embryonic germ layers. Mol Med 6(2):88–95

129. Carr AJ, Smart MJ, Ramsden CM, Powner MB, da Cruz L, Coffey PJ (2013) Development of human embryonic stem cell therapies for age-related macular degeneration. Trends Neurosci 36(7):385–395

130. Idelson M, Alper R, Obolensky A, Ben-Shushan E, Hemo I, Yachimovich-Cohen N, Khaner H, Smith Y, Wiser O, Gropp M, Cohen MA, Even-Ram S, Berman-Zaken Y, Matzrafi L, Rechavi G, Banin E, Reubinoff B (2009) Directed differentiation of human embryonic stem cells into functional retinal pigment epithelium cells. Cell Stem Cell 5(4):396–408

131. Meyer JS, Howden SE, Wallace KA, Verhoeven AD, Wright LS, Capowski EE, Pinilla I, Martin JM, Tian S, Stewart R, Pattnaik B, Thomson JA, Gamm DM (2011) Optic vesicle-like structures derived from human pluripotent stem cells facilitate a customized approach to retinal disease treatment. Stem Cells 29(8):1206–1218

132. Reichman S, Slembrouck A, Gagliardi G, Chaffiol A, Terray A, Nanteau C, Potey A, Belle M, Rabesandratana O, Duebel J, Orieux G, Nandrot EF, Sahel JA, Goureau O (2017) Generation of storable retinal organoids and retinal pigmented epithelium from adherent human iPS cells in xeno-free and feeder-free conditions. Stem Cells 35(5):1176–1188

133. Reichman S, Terray A, Slembrouck A, Nanteau C, Orieux G, Habeler W, Nandrot EF, Sahel JA, Monville C, Goureau O (2014) From confluent human iPS cells to self-forming neural retina and retinal pigmented epithelium. Proc Natl Acad Sci U S A 111(23):8518–8523

134. Crombie DE, Daniszewski M, Liang HH, Kulkarni T, Li F, Lidgerwood GE, Conquest A, Hernandez D, Hung SS, Gill KP, De Smit E, Kearns LS, Clarke L, Sluch VM, Chamling X, Zack DJ, Wong RCB, Hewitt AW, Pebay A (2017) Development of a modular automated system for maintenance and differentiation of adherent human pluripotent stem cells. SLAS Discov 22(8):1016–1025

135. Schwartz SD, Hubschman JP, Heilwell G, Franco-Cardenas V, Pan CK, Ostrick RM, Mickunas E, Gay R, Klimanskaya I, Lanza R (2012) Embryonic stem cell trials for macular degeneration: a preliminary report. Lancet 379(9817):713–720

136. Mandai M, Watanabe A, Kurimoto Y, Hirami Y, Morinaga C, Daimon T, Fujihara M, Akimaru H, Sakai N, Shibata Y, Terada M, Nomiya Y, Tanishima S, Nakamura M, Kamao H, Sugita S, Onishi A, Ito T, Fujita K, Kawamata S, Go MJ, Shinohara C, Hata KI, Sawada M, Yamamoto M, Ohta S, Ohara Y, Yoshida K, Kuwahara J, Kitano Y, Amano N, Umekage M, Kitaoka F, Tanaka A, Okada C, Takasu N, Ogawa S, Yamanaka S, Takahashi M (2017) Autologous induced stem-cell-derived retinal cells for macular degeneration. N Engl J Med 376(11):1038–1046

137. Xu RM, Carmel G, Kuret J, Cheng X (1996) Structural basis for selectivity of the isoquinoline sulfonamide family of protein kinase inhibitors. Proc Natl Acad Sci U S A 93(13):6308–6313

138. Ben M'Barek K, Habeler W, Plancheron A, Jarraya M, Regent F, Terray A, Yang Y, Chatrousse L, Domingues S, Masson Y, Sahel JA, Peschanski M, Goureau O, Monville C (2017) Human ESC-derived retinal epithelial cell sheets potentiate rescue of photoreceptor cell loss in rats with retinal degeneration. Sci Transl Med 9(421):eaai7471

139. Lustremant C, Habeler W, Plancheron A, Goureau O, Grenot L, de la Grange P, Audo I, Nandrot EF, Monville C (2013) Human induced pluripotent stem cells as a tool to model a form of Leber congenital amaurosis. Cell Reprogram 15(3):233–246

140. Schwartz SD, Regillo CD, Lam BL, Eliott D, Rosenfeld PJ, Gregori NZ, Hubschman JP, Davis JL, Heilwell G, Spirn M, Maguire J, Gay R, Bateman J, Ostrick RM, Morris D, Vincent M, Anglade E, Del Priore LV, Lanza R (2015) Human embryonic stem cell-derived retinal pigment epithelium in patients with age-related macular degeneration and Stargardt's macular dystrophy: follow-up of two open-label phase 1/2 studies. Lancet 385(9967):509–516

141. Andrews PW, Baker D, Benvinisty N, Miranda B, Bruce K, Brustle O, Choi M, Choi YM, Crook JM, de Sousa PA, Dvorak P, Freund C, Firpo M, Furue MK, Gokhale P, Ha HY, Han E, Haupt S, Healy L, Hei DJ, Hovatta O, Hunt C, Hwang SM, Inamdar MS, Isasi RM, Jaconi M, Jekerle V, Kamthorn P, Kibbey MC, Knezevic I, Knowles BB, Koo SK, Laabi Y, Leopoldo L, Liu P, Lomax GP, Loring JF, Ludwig TE, Montgomery K, Mummery C, Nagy A, Nakamura Y, Nakatsuji N, Oh S, Oh SK, Otonkoski T, Pera M, Peschanski M, Pranke P, Rajala KM,

Rao M, Ruttachuk R, Reubinoff B, Ricco L, Rooke H, Sipp D, Stacey GN, Suemori H, Takahashi TA, Takada K, Talib S, Tannenbaum S, Yuan BZ, Zeng F, Zhou Q (2015) Points to consider in the development of seed stocks of pluripotent stem cells for clinical applications: International Stem Cell Banking Initiative (ISCBI). Regen Med 10(2 Suppl):1–44

142. Kuroda T, Yasuda S, Kusakawa S, Hirata N, Kanda Y, Suzuki K, Takahashi M, Nishikawa S, Kawamata S, Sato Y (2012) Highly sensitive in vitro methods for detection of residual undifferentiated cells in retinal pigment epithelial cells derived from human iPS cells. PLoS One 7(5):e37342

143. Kanemura H, Go MJ, Shikamura M, Nishishita N, Sakai N, Kamao H, Mandai M, Morinaga C, Takahashi M, Kawamata S (2014) Tumorigenicity studies of induced pluripotent stem cell (iPSC)-derived retinal pigment epithelium (RPE) for the treatment of age-related macular degeneration. PLoS One 9(1):e85336

144. Bai Q, Ramirez JM, Becker F, Pantesco V, Lavabre-Bertrand T, Hovatta O, Lemaitre JM, Pellestor F, De Vos J (2015) Temporal analysis of genome alterations induced by single-cell passaging in human embryonic stem cells. Stem Cells Dev 24(5):653–662

145. Weissbein U, Benvenisty N, Ben-David U (2014) Quality control: genome maintenance in pluripotent stem cells. J Cell Biol 204(2):153–163

146. Garber K (2015) RIKEN suspends first clinical trial involving induced pluripotent stem cells. Nat Biotechnol 33(9):890–891

147. Georgiadis A, Tschernutter M, Bainbridge JW, Balaggan KS, Mowat F, West EL, Munro PM, Thrasher AJ, Matter K, Balda MS, Ali RR (2010) The tight junction associated signalling proteins ZO-1 and ZONAB regulate retinal pigment epithelium homeostasis in mice. PLoS One 5(12):e15730

148. Kannan R, Zhang N, Sreekumar PG, Spee CK, Rodriguez A, Barron E, Hinton DR (2006) Stimulation of apical and basolateral VEGF-A and VEGF-C secretion by oxidative stress in polarized retinal pigment epithelial cells. Mol Vis 12:1649–1659

149. Nazari H, Zhang L, Zhu D, Chader GJ, Falabella P, Stefanini F, Rowland T, Clegg DO, Kashani AH, Hinton DR, Humayun MS (2015) Stem cell based therapies for age-related macular degeneration: the promises and the challenges. Prog Retin Eye Res 48:1–39

150. Capeans C, Pineiro A, Pardo M, Sueiro-Lopez C, Blanco MJ, Dominguez F, Sanchez-Salorio M (2003) Amniotic membrane as support for human retinal pigment epithelium (RPE) cell growth. Acta Ophthalmol Scand 81(3):271–277

151. Niknejad H, Peirovi H, Jorjani M, Ahmadiani A, Ghanavi J, Seifalian AM (2008) Properties of the amniotic membrane for potential use in tissue engineering. Eur Cell Mater 15:88–99

152. Adds PJ, Hunt CJ, Dart JK (2001) Amniotic membrane grafts, "fresh" or frozen? A clinical and in vitro comparison. Br J Ophthalmol 85(8):905–907

153. Kador KE, Goldberg JL (2012) Scaffolds and stem cells: delivery of cell transplants for retinal degenerations. Expert Rev Ophthalmol 7(5):459–470

154. Hu Y, Liu L, Lu B, Zhu D, Ribeiro R, Diniz B, Thomas PB, Ahuja AK, Hinton DR, Tai YC, Hikita ST, Johnson LV, Clegg DO, Thomas BB, Humayun MS (2012) A novel approach for subretinal implantation of ultrathin substrates containing stem cell-derived retinal pigment epithelium monolayer. Ophthalmic Res 48(4):186–191

155. da Cruz L, Fynes K, Georgiadis O, Kerby J, Luo YH, Ahmado A, Vernon A, Daniels JT, Nommiste B, Hasan SM, Gooljar SB, Carr AF, Vugler A, Ramsden CM, Bictash M, Fenster M, Steer J, Harbinson T, Wilbrey A, Tufail A, Feng G, Whitlock M, Robson AG, Holder GE, Sagoo MS, Loudon PT, Whiting P, Coffey PJ (2018) Phase 1 clinical study of an embryonic stem cell-derived retinal pigment epithelium patch in age-related macular degeneration. Nat Biotechnol 36(4):328–337

156. Seiler MJ, Aramant RB (2012) Cell replacement and visual restoration by retinal sheet transplants. Prog Retin Eye Res 31(6):661–687

157. Bharti K, Rao M, Hull SC, Stroncek D, Brooks BP, Feigal E, van Meurs JC, Huang CA, Miller SS (2014) Developing cellular therapies for retinal degenerative diseases. Invest Ophthalmol Vis Sci 55(2):1191–1202

158. Song WK, Park KM, Kim HJ, Lee JH, Choi J, Chong SY, Shim SH, Del Priore LV, Lanza R (2015) Treatment of macular degeneration using embryonic stem cell-derived retinal pigment epithelium: preliminary results in Asian patients. Stem Cell Rep 4(5):860–872
159. Ramsden CM, Powner MB, Carr AJ, Smart MJ, da Cruz L, Coffey PJ (2013) Stem cells in retinal regeneration: past, present and future. Development 140(12):2576–2585
160. Brandl C, Grassmann F, Riolfi J, Weber BH (2015) Tapping stem cells to target AMD: challenges and prospects. J Clin Med 4(2):282–303
161. Kuriyan AE, Albini TA, Townsend JH, Rodriguez M, Pandya HK, Leonard RE II, Parrott MB, Rosenfeld PJ, Flynn HW Jr, Goldberg JL (2017) Vision loss after intravitreal injection of autologous "stem cells" for AMD. N Engl J Med 376(11):1047–1053
162. Forrester JV (2009) Privilege revisited: an evaluation of the eye's defence mechanisms. Eye (Lond) 23(4):756–766
163. Streilein JW, Ma N, Wenkel H, Ng TF, Zamiri P (2002) Immunobiology and privilege of neuronal retina and pigment epithelium transplants. Vision Res 42(4):487–495
164. Sauve Y, Klassen H, Whiteley SJ, Lund RD (1998) Visual field loss in RCS rats and the effect of RPE cell transplantation. Exp Neurol 152(2):243–250
165. Araki R, Uda M, Hoki Y, Sunayama M, Nakamura M, Ando S, Sugiura M, Ideno H, Shimada A, Nifuji A, Abe M (2013) Negligible immunogenicity of terminally differentiated cells derived from induced pluripotent or embryonic stem cells. Nature 494(7435):100–104
166. Drukker M, Katz G, Urbach A, Schuldiner M, Markel G, Itskovitz-Eldor J, Reubinoff B, Mandelboim O, Benvenisty N (2002) Characterization of the expression of MHC proteins in human embryonic stem cells. Proc Natl Acad Sci U S A 99(15):9864–9869
167. Wu KH, Wu HP, Chan CK, Hwang SM, Peng CT, Chao YH (2013) The role of mesenchymal stem cells in hematopoietic stem cell transplantation: from bench to bedsides. Cell Transplant 22(4):723–729
168. Sohn EH, Jiao C, Kaalberg E, Cranston C, Mullins RF, Stone EM, Tucker BA (2015) Allogenic iPSC-derived RPE cell transplants induce immune response in pigs: a pilot study. Sci Rep 5:11791
169. Xian B, Huang B (2015) The immune response of stem cells in subretinal transplantation. Stem Cell Res Ther 6:161
170. Zamiri P, Sugita S, Streilein JW (2007) Immunosuppressive properties of the pigmented epithelial cells and the subretinal space. Chem Immunol Allergy 92:86–93
171. Wenkel H, Streilein JW (1998) Analysis of immune deviation elicited by antigens injected into the subretinal space. Invest Ophthalmol Vis Sci 39(10):1823–1834
172. Sugita S, Futagami Y, Smith SB, Naggar H, Mochizuki M (2006) Retinal and ciliary body pigment epithelium suppress activation of T lymphocytes via transforming growth factor beta. Exp Eye Res 83(6):1459–1471
173. Zamiri P, Masli S, Streilein JW, Taylor AW (2006) Pigment epithelial growth factor suppresses inflammation by modulating macrophage activation. Invest Ophthalmol Vis Sci 47(9):3912–3918
174. Wenkel H, Streilein JW (2000) Evidence that retinal pigment epithelium functions as an immune-privileged tissue. Invest Ophthalmol Vis Sci 41(11):3467–3473
175. Hirsch L, Nazari H, Sreekumar PG, Kannan R, Dustin L, Zhu D, Barron E, Hinton DR (2015) TGF-beta2 secretion from RPE decreases with polarization and becomes apically oriented. Cytokine 71(2):394–396
176. Zamiri P, Masli S, Kitaichi N, Taylor AW, Streilein JW (2005) Thrombospondin plays a vital role in the immune privilege of the eye. Invest Ophthalmol Vis Sci 46(3):908–919
177. Enzmann V, Stadler M, Wiedemann P, Kohen L (1998) In-vitro methods to decrease MHC class II-positive cells in retinal pigment epithelium cell grafts. Ocul Immunol Inflamm 6(3):145–153
178. Horie S, Sugita S, Futagami Y, Yamada Y, Mochizuki M (2010) Human retinal pigment epithelium-induced CD4+CD25+ regulatory T cells suppress activation of intraocular effector T cells. Clin Immunol 136(1):83–95

179. Sugita S, Horie S, Nakamura O, Maruyama K, Takase H, Usui Y, Takeuchi M, Ishidoh K, Koike M, Uchiyama Y, Peters C, Yamamoto Y, Mochizuki M (2009) Acquisition of T regulatory function in cathepsin L-inhibited T cells by eye-derived CTLA-2alpha during inflammatory conditions. J Immunol 183(8):5013–5022

180. Sugita S, Horie S, Yamada Y, Mochizuki M (2010) Inhibition of B-cell activation by retinal pigment epithelium. Invest Ophthalmol Vis Sci 51(11):5783–5788

181. Sugita S, Usui Y, Horie S, Futagami Y, Yamada Y, Ma J, Kezuka T, Hamada H, Usui T, Mochizuki M, Yamagami S (2009) Human corneal endothelial cells expressing programmed death-ligand 1 (PD-L1) suppress PD-1+ T helper 1 cells by a contact-dependent mechanism. Invest Ophthalmol Vis Sci 50(1):263–272

182. Osusky R, Dorio RJ, Arora YK, Ryan SJ, Walker SM (1997) MHC class II positive retinal pigment epithelial (RPE) cells can function as antigen-presenting cells for microbial superantigen. Ocul Immunol Inflamm 5(1):43–50

183. Imai A, Sugita S, Kawazoe Y, Horie S, Yamada Y, Keino H, Maruyama K, Mochizuki M (2012) Immunosuppressive properties of regulatory T cells generated by incubation of peripheral blood mononuclear cells with supernatants of human RPE cells. Invest Ophthalmol Vis Sci 53(11):7299–7309

184. Futagami Y, Sugita S, Vega J, Ishida K, Takase H, Maruyama K, Aburatani H, Mochizuki M (2007) Role of thrombospondin-1 in T cell response to ocular pigment epithelial cells. J Immunol 178(11):6994–7005

185. Diniz B, Thomas P, Thomas B, Ribeiro R, Hu Y, Brant R, Ahuja A, Zhu D, Liu L, Koss M, Maia M, Chader G, Hinton DR, Humayun MS (2013) Subretinal implantation of retinal pigment epithelial cells derived from human embryonic stem cells: improved survival when implanted as a monolayer. Invest Ophthalmol Vis Sci 54(7):5087–5096

186. Hambright D, Park KY, Brooks M, McKay R, Swaroop A, Nasonkin IO (2012) Long-term survival and differentiation of retinal neurons derived from human embryonic stem cell lines in un-immunosuppressed mouse retina. Mol Vis 18:920–936

187. Lu B, Malcuit C, Wang S, Girman S, Francis P, Lemieux L, Lanza R, Lund R (2009) Long-term safety and function of RPE from human embryonic stem cells in preclinical models of macular degeneration. Stem Cells 27(9):2126–2135

188. Hewitt Z, Priddle H, Thomson AJ, Wojtacha D, McWhir J (2007) Ablation of undifferentiated human embryonic stem cells: exploiting innate immunity against the Gal alpha1-3Galbeta1-4GlcNAc-R (alpha-Gal) epitope. Stem Cells 25(1):10–18

189. Sugita S, Makabe K, Fujii S, Iwasaki Y, Kamao H, Shiina T, Ogasawara K, Takahashi M (2017) Detection of retinal pigment epithelium-specific antibody in iPSC-derived retinal pigment epithelium transplantation models. Stem Cell Rep 9(5):1501–1515

190. Sonoda S, Sreekumar PG, Kase S, Spee C, Ryan SJ, Kannan R, Hinton DR (2010) Attainment of polarity promotes growth factor secretion by retinal pigment epithelial cells: relevance to age-related macular degeneration. Aging (Albany NY) 2(1):28–42

191. Hsiung J, Zhu D, Hinton DR (2015) Polarized human embryonic stem cell-derived retinal pigment epithelial cell monolayers have higher resistance to oxidative stress-induced cell death than nonpolarized cultures. Stem Cells Transl Med 4(1):10–20

192. Kawazoe Y, Sugita S, Keino H, Yamada Y, Imai A, Horie S, Mochizuki M (2012) Retinoic acid from retinal pigment epithelium induces T regulatory cells. Exp Eye Res 94(1):32–40

193. Sugita S, Horie S, Nakamura O, Futagami Y, Takase H, Keino H, Aburatani H, Katunuma N, Ishidoh K, Yamamoto Y, Mochizuki M (2008) Retinal pigment epithelium-derived CTLA-2alpha induces TGFbeta-producing T regulatory cells. J Immunol 181(11):7525–7536

194. Sugita S, Kamao H, Iwasaki Y, Okamoto S, Hashiguchi T, Iseki K, Hayashi N, Mandai M, Takahashi M (2015) Inhibition of T-cell activation by retinal pigment epithelial cells derived from induced pluripotent stem cells. Invest Ophthalmol Vis Sci 56(2):1051–1062

195. Ferguson TA, Griffith TS (2007) The role of Fas ligand and TNF-related apoptosis-inducing ligand (TRAIL) in the ocular immune response. Chem Immunol Allergy 92:140–154

196. Sugita S, Kawazoe Y, Imai A, Usui Y, Takahashi M, Mochizuki M (2013) Suppression of IL-22-producing T helper 22 cells by RPE cells via PD-L1/PD-1 interactions. Invest Ophthalmol Vis Sci 54(10):6926–6933
197. Zhang X, Bok D (1998) Transplantation of retinal pigment epithelial cells and immune response in the subretinal space. Invest Ophthalmol Vis Sci 39(6):1021–1027
198. Bhatt NS, Newsome DA, Fenech T, Hessburg TP, Diamond JG, Miceli MV, Kratz KE, Oliver PD (1994) Experimental transplantation of human retinal pigment epithelial cells on collagen substrates. Am J Ophthalmol 117(2):214–221
199. Crafoord S, Algvere PV, Kopp ED, Seregard S (2000) Cyclosporine treatment of RPE allografts in the rabbit subretinal space. Acta Ophthalmol Scand 78(2):122–129
200. Ahmad ZM, Hughes BA, Abrams GW, Mahmoud TH (2012) Combined posterior chamber intraocular lens, vitrectomy, Retisert, and pars plana tube in noninfectious uveitis. Arch Ophthalmol 130(7):908–913
201. Tomkins-Netzer O, Taylor SR, Bar A, Lula A, Yaganti S, Talat L, Lightman S (2014) Treatment with repeat dexamethasone implants results in long-term disease control in eyes with noninfectious uveitis. Ophthalmology 121(8):1649–1654
202. Grafft CA, Cornell LD, Gloor JM, Cosio FG, Gandhi MJ, Dean PG, Stegall MD, Amer H (2010) Antibody-mediated rejection following transplantation from an HLA-identical sibling. Nephrol Dial Transplant 25(1):307–310
203. Taylor CJ, Bolton EM, Pocock S, Sharples LD, Pedersen RA, Bradley JA (2005) Banking on human embryonic stem cells: estimating the number of donor cell lines needed for HLA matching. Lancet 366(9502):2019–2025
204. Copelan EA (2006) Hematopoietic stem-cell transplantation. N Engl J Med 354(17): 1813–1826
205. de Rham C, Villard J (2014) Potential and limitation of HLA-based banking of human pluripotent stem cells for cell therapy. J Immunol Res 2014:518135
206. Okita K, Nagata N, Yamanaka S (2011) Immunogenicity of induced pluripotent stem cells. Circ Res 109(7):720–721
207. Opelz G, Dohler B (2010) Impact of HLA mismatching on incidence of posttransplant non-hodgkin lymphoma after kidney transplantation. Transplantation 89(5):567–572
208. Sugita S, Iwasaki Y, Makabe K, Kimura T, Futagami T, Suegami S, Takahashi M (2016) Lack of T cell response to iPSC-derived retinal pigment epithelial cells from HLA homozygous donors. Stem Cell Rep 7(4):619–634
209. Sugita S, Iwasaki Y, Makabe K, Kamao H, Mandai M, Shiina T, Ogasawara K, Hirami Y, Kurimoto Y, Takahashi M (2016) Successful transplantation of retinal pigment epithelial cells from MHC homozygote iPSCs in MHC-matched models. Stem Cell Rep 7(4):635–648
210. Nakajima F, Tokunaga K, Nakatsuji N (2007) Human leukocyte antigen matching estimations in a hypothetical bank of human embryonic stem cell lines in the Japanese population for use in cell transplantation therapy. Stem Cells 25(4):983–985
211. Gourraud PA, Gilson L, Girard M, Peschanski M (2012) The role of human leukocyte antigen matching in the development of multiethnic "haplobank" of induced pluripotent stem cell lines. Stem Cells 30(2):180–186

Chapter 4
Immunological Considerations for Retinal Stem Cell Therapy

Joshua Kramer, Kathleen R. Chirco, and Deepak A. Lamba

Abstract There is an increasing effort toward generating replacement cells for neuro-nal application due to the nonregenerative nature of these tissues. While much prog-ress has been made toward developing methodologies to generate these cells, there have been limited improvements in functional restoration. Some of these are linked to the degenerative and often nonreceptive microenvironment that the new cells need to integrate into. In this chapter, we will focus on the status and role of the immune microenvironment of the retina during homeostasis and disease states. We will review changes in both innate and adaptive immunity as well as the role of immune rejection in stem cell replacement therapies. The chapter will end with a discussion of immune-modulatory strategies that have helped to ameliorate these effects and could poten-tially improve functional outcome for cell replacement therapies for the eye.

Keywords Immune modulation · Cell replacement · Microglia · Macrophages · Muller glia · Complement system · Immune rejection · Inflammasome

4.1 Introduction

Much progress has been made in the field of cell transplantation in the retina, with various groups demonstrating improved cell survival and integration. For individu-als with chronic degenerative conditions, this progress in the field of cell transplan-tation provides promise for future success in replacement of dead or dying cells

J. Kramer · K. R. Chirco
Department of Ophthalmology, University of California San Francisco,
San Francisco, CA, USA

D. A. Lamba (✉)
Buck Institute for Research on Aging, Novato, CA, USA

Department of Ophthalmology, University of California San Francisco,
San Francisco, CA, USA
e-mail: deepak.lamba@ucsf.edu

© Springer Nature Switzerland AG 2019
K. Bharti (ed.), *Pluripotent Stem Cells in Eye Disease Therapy*,
Advances in Experimental Medicine and Biology 1186,
https://doi.org/10.1007/978-3-030-28471-8_4

with new healthy ones in tissues with no regenerative capacity. Our lab and many others have been working toward developing technologies to generate replacement cells for the retina, including photoreceptors, retinal pigment epithelium (RPE), and retinal ganglion cells [1–7]. While there is currently a major effort to improve protocols for the generatation of retinal cell types, and to optimize delivery methods and substrates, improvements in integration efficiencies can still be made, especially with regard to the microenvironment these cells require for survival and integration. One key factor to consider in cell transplantation is the role of the immune system in tissue damage, recovery, and donor cell rejection. Here, we review the current work being done to understand immune modulation as it relates to cell replacement therapies in the retina.

4.2   Immune System Development and Role in Ocular Tissue Maintenance

Immune protection within the retina is carried out through cell-based immune systems, including the innate and adaptive immune systems, as well as the protein-based complement system. Although the majority of components and activities within these systems are identical whether they are acting inside or outside of the eye, there are a few distinct differences that are inherent to the central nervous system (CNS), or to the retina specifically. For example, acting as an initial line of defense, the CNS is protected from harmful pathogens and other toxic substances via the blood–brain barrier (BBB) and blood–retinal barrier (BRB). An array of cell types are required for the formation of these barriers, including endothelial cells, pericytes, basement membrane, retinal pigment epithelium (RPE, in the case of the retina), glia (astrocytes and/or Müller glia), and microglia. The BRB, which consists of an inner and outer barrier, allows for precise regulation of ions, oxygen, nutrients, and cells that pass between the blood and the retina. More specifically, the outer BRB refers to the tight junctions between RPE cells and provides a barrier between the choroid and the outer neural retina. The inner BRB, which more closely resembles the characteristics of the BBB, surrounds the inner retinal microvasculature. Despite its important role in defense against invading pathogens and harmful substances, the existence of the BRB is not, by itself, sufficient to protect the retina from extensive tissue injury or disease.

Beyond these barriers, the microenvironment of the eye plays an immunosuppressive role. Interstitial transforming growth factor beta (TGF-β) is ubiquitous within the vitreous of the adult eye, acting to inhibit immune activation [8–10]. TGF-β works by blocking synthesis of p21 and p15, proteins that block cyclin-CDK [11, 12]. In vitro studies have shown that TGF-β produced by RPE cells can drive CD4+ T cells to become Tregs in the presence of CTLA-2α [13]. This has been confirmed in vivo using Foxp3-GFP mice that mark Tregs following local antigen recognition [14]. While the above-mentioned barriers to cell rejection may promote immune-privilege in the retina, these barriers do get compromised with both aging and disease.

Autoimmune and immune-mediated diseases of the eye occur at high frequency suggesting that this barrier can be easily broken. Therefore, the retina must rely on other arms of the immune system to remain protected and overcome disease.

4.2.1 Innate Immune System

Cells of the innate immune system are crucial in mediating the response to injury and aiding in recovery. Neutrophils, which exhibit both pro- and anti-inflammatory functions, are the first responders to damaged tissue. In addition to their ability to metabolize hydrogen peroxide via the release of myeloperoxidase at the injury site, they also play a key role in the recruitment of monocytes to the tissue for rapid clearance of dying cells and debris. Another early responder to tissue injury is the mast cell. Mast cells are extremely heterogeneous cells that have the potential to secrete eosinophil- and monocyte-specific chemoattractants, such as eosinophil chemotactic factor, TNF-α, and IL-4. Mast cells also release various effector molecules that can promote angiogenesis and epithelial cell growth, encourage glial scarring, and increase inflammation, depending on the requirements of the tissue (reviewed in [15]).

In response to specific chemokines and cytokines, monocyte-derived macrophages are recruited from the vasculature to sites of damage and inflammation. Macrophages are present at all stages of the tissue repair process and play an integral role in both amplification and mitigation of the immune response to tissue damage [16–18]. When macrophages are primed by injury-driven interferon-γ (IFN-γ), they become pro-inflammatory M1 macrophages [19]. M1 macrophages are characterized by the production of TNF-α and interleukin-12 (IL-12) and are associated with tissue damage. On the other hand, macrophages activated in the presence of IL-4 differentiate into the nonclassical, wound-healing M2-type, marked by production of the immunosuppressive cytokine IL-10 [20, 21]. In contrast to the M1 subtype, M2 macrophages are involved in tissue remodeling (Fig. 4.1). Monocyte-derived macrophages phagocytose dying cells and tissue debris to aid in their rapid removal. Macrophages can also activate and/or propagate the innate and adaptive immune systems to aid in tissue repair.

Unique to the CNS, microglia are long-lived, self-sustaining phagocytic mononuclear cells that perform similar functions to monocyte-derived macrophages in other tissues [17]. As the principal resident innate immune cell within the retina, microglia reside in a horizontal lattice within the inner and outer plexiform layers, which coincides with retinal synaptogenesis [22–24]. Microglia have the ability to trigger apoptosis in unfit cells, phagocytose dead or dying cells and debris, and constitutively monitor the function of retinal synapses and surrounding extracellular matrix [25–27]. Along with the traditional role of immune defense, microglia play crucial roles in development, homeostasis, and neurodegeneration.

Early microglia express markers such as CD68, CD45, F4/80, and isolectin, along with known inflammatory markers not usually present in healthy adult microglia, including iNOS, superoxide ions, and TNF-α [28]. Expression of these markers may allow for microglial control of neuronal programmed cell death during retinal

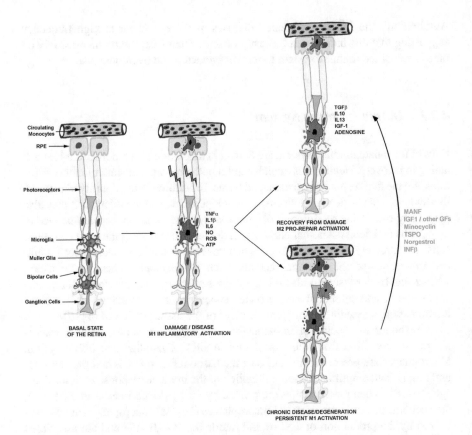

Fig. 4.1 Microglia/Macrophage activation in retinal disorders. Cartoon shows changes in resident microglia and migrating macrophages during acute and chronic degenerative conditions. Damage results in inflammatory activation of resident microglia and invasion of circulating macrophages into the retina. These cells transition to a prorepair M2 state to allow the tissue to heal. However, chronic degenerative disorders lead to persistent inflammatory activation that will require immune modulators including MANF to promote repair

development as part of the normal pruning process, which is likely made easier by microglial proximity to developing retinal synapses [29–32]. Additionally, developing microglia secrete growth factors for both neurons and vasculature [33–35]; therefore, neovascularization may, at least in part, be dictated by developing microglia via secreted trophic factors. Once matured to adulthood, the uniformly spaced microglia of the inner and outer plexiform layer transition from an ameboid into a ramified resting state use their processes to clear metabolic waste and to continually survey nearby synapses and vasculature [30]. In the healthy retina, microglia exist as a self-sustaining immune cell population with low turnover, with some individual microglia persisting for the lifetime of the organism [36, 37].

Acting as one of the first lines of defense against tissue damage or dysfunction, Müller cells, or Müller glia, are sensitive to changes that occur within the retina, especially those resulting from a disease state. The Müller cell is a retina-specific glial cell that spans the neural retina from the inner limiting membrane to the outer limiting membrane. Müller glia are generated from the terminal division of retinal progenitor cells and are one of the last cell types born in the retina [38–40]. Their main role is to provide structural and functional support to retinal neurons, including neurotransmitter uptake and recycling, regulation of water and ion levels, waste removal, and recycling of cone photopigments, in addition to contributing to the outer BRB (reviewed in [39]). Acute disruptions to normal retinal function result in a protective form of Müller cell gliosis, which is primarily characterized by increased glial cell rigidity (via upregulation of GFAP and vimentin intermediate filament proteins) and an increased number of processes. Gliotic Müller cells can also produce antioxidants to protect the neural retina during stress (reviewed in [26]). However, in chronic conditions (e.g., retinal degenerative diseases), Müller glia upregulate proinflammatory molecules to increase inflammation (discussed below).

4.2.2 Adaptive Immune System

The adaptive immune system comprises two major types of lymphocytes: T cells and B cells. Although adaptive immunity is typically activated in response to an invading pathogen, recent work suggests an alternative role for T cells in maintaining normal tissue health. For example, in response to a foreign antigen, both effector T cells, which ramp up the immune response, and regulatory T cells (T_{reg}), which help to dampen the immune response, are utilized in a tightly controlled balance [41]. However, in addition to their role in adaptive immunity, recent studies have implicated a subset of T_{reg} cells in healthy tissue repair mechanisms [15, 42]. The absence of T_{reg} cells has been shown to impair muscle tissue repair [43, 44], and importantly, T_{reg} cells demonstrated an ability to regulate macrophage and monocyte numbers within the damaged tissue and promote an M2 phenotype. T_{reg} cells also negatively regulate the number of effector T cells present in damaged tissue, which aids in tissue repair [43].

4.2.3 Complement System

The complement system is made up of over 30 serum- and tissue-associated proteins that are traditionally viewed as mediators in immune response induction or regulation [45]. Three major pathways make up the complement system, each of which is triggered by a unique initiation event, followed by a cascade of protein cleavage events. Key components that are common to all three complement pathways include C3, the anaphylatoxin C3a, the formation of a C3 convertase (either C4b2a or C3bBb), C5, the anaphylatoxin C5a, the formation of C5 convertases (either C4b2a3b or C3bBb3b), and the formation of the terminal complement complex, also known as the membrane attack complex (MAC; C5b9). Although their exact initiation events differ, the primary role of all three complement pathways throughout the body is to help clear pathogenic cells as well as dead or dying host cells. To carry out this role, the complement system can (1) opsonize harmful antigens (via C3b activity), (2) trigger or amplify the inflammatory response by attracting marcophages and neutrophils to the site of tissue damage or infection, and/or (3) disrupt pathogenic cell membranes by depositing MAC onto the surface of the organism. Many regulators of complement activation act in concert with pro-inflammatory complement proteins (e.g., factor H, protectin, and decay-accelerating factor), as complement regulation is critical to prevent unnecessary activation or propagation of an immune response.

In addition to participating in an immune response, studies performed in the embryonic chick retina revealed the anaphylatoxin, C3a, as an inducer of retinal regeneration in vivo [46]. Injection of recombinant C3a was found to be sufficient to trigger regeneration of the chick retina from stem cells located in the ciliary margin. C3a-mediated regeneration utilizes the MAPK/STAT3 signaling pathway and requires the phosphorylation of STAT3 at serine 727 [46]. This work presents an intriguing hypothesis that complement proteins could aid in retinal regeneration (reviewed by Hawksworth et al. [47]). Furthermore, extracellular MAC levels in the choroid (behind the RPE layer) increase with age in healthy human eyes, beginning as early as 5 years of age [48]. This phenomenon could suggest a beneficial role for the MAC in the eye, although more work is required to confirm this. While these data provide promising insight into the complement system as a tool for improving tissue health and regeneration, care must be taken to maintain the balance between complement activation and regulation, and tissue-specific differences in complement-mediated regeneration must be taken into consideration.

4.3 Damage- and Disease-Associated Inflammation in the Eye

Degenerative diseases of the retina often coincide with or directly cause inflammation. When it comes to treating such diseases, especially in the case of cell transplantation, the disease-mediated inflammatory environment in the recipient tissue must be carefully considered.

4.3.1 Microglia/Macrophage Activation

As stated above, the mononuclear immune cells including microglia and macrophages are critical innate immune responders to any retinal damage. The retinal inflammation process is moderated by microglia, which upon sensing the injury becomes activated M1 microglia [49]. Intracellular miRNA-155 expression has been shown to be important to keep them in an M1 state [50]. These M1 microglia migrate to the source of injury, proliferate, and phagocytose debris, working with nearby astrocytes to release inflammatory cytokines and chemokines such as TNF-α, IL-1β, IL-6, IL-8, glutamate, reactive oxygen species, and nitric oxide, all of which can result in the death of nearby susceptible neurons [51]. Upon activation, microglia have also been shown to initiate cross-talk with nearby Müller glia, which can induce a gliotic response [52].

As the initial damage-signaling attenuates, M1 microglia transition to an M2 anti-inflammatory phenotype or undergo apoptosis to make way for new M2-polarized microglia. The M2 transition process normally occurs by activation of intrinsic molecular switches such as STAT6, IFR-4, PPARγ, and C/EBPB, along with miRNA-124 [50]. There are three distinct M2 expression profiles: M2a, M2b, and M2c. Beginning chronologically, M2a microglia primarily participate in inflammation inhibition and secretion of anti-inflammatory cytokines (IL-4, IL-13, and IL-10; [20, 21, 53]). Once this process is initiated, M2c microglia initiate restoration and repair of damaged surviving cells by (1) secreting neurotrophic factors such as the calcium binding protein oncomodulin [54] and IGF-1 to induce neurogenesis [55], (2) secreting vascular endothelial growth factor (VEGF) to stimulate angiogenesis [56, 57], and (3) promoting oligodendrogenesis and remyelination [58, 59]. M2b microglia, which form via the final microglia subdivision, have the ability to activate complement, are involved in the memory immune response, and can help to stimulate or reduce inflammation as necessary [60]. At the conclusion of this polarization process, microglia return to a ramified resting state.

The M2 transition typically occurs under states of acute injury or stress and can help return the retina to homeostasis. However, if the insult persists, as in the case of inherited retinal degenerations or age-related macular degeneration, the inflammatory response becomes increasingly toxic to the cells and surrounding tissue within the microenvironment [49]. Furthermore, our innate immune system becomes increasingly dysregulated with increasing age, which has been made evident by a persistent M1-type inflammatory response in numerous aged tissues [61]. These overreactive M1 microglia/macrophages release proinflammatory and cytotoxic factors including TNF-α and IL-1β. Additionally, it has been described recently that this can lead to complement pathway dysfunction [62]. Activated microglia can cause upregulation of complement activators C3, CFB, C1q, and C5AR1 and downregulating complement inhibitors CFH, CFI, CD46, and CD93. These observations are supported by additional studies showing reduced light damage-induced recruitment of microglia to the outer retina and the subretinal space in mice lacking functional C5aR [63].

As previously mentioned, Müller cells typically undergo reactive gliosis in response to damage, infection, or other inflammatory conditions within the retina. Chronic gliosis causes damage through the production of proinflammatory cytokines and nitric oxide. Additionally, Hippert and colleagues recently reported an increase in Müller glia reactivity with increasing age in control mice. This increase in gliosis was accompanied by an increase in chondroitin sulfate proteoglycan (CSPG) deposition throughout the neural retina, including the photoreceptor inner and outer segment layer [64]. Elevated levels of CSPGs in the retina have been shown to inhibit axon regeneration and may represent a significant concern that needs to be addressed in cell replacement therapies [65, 66]. Recently, IL-33 has been observed in the nucleus of Müller cells of the healthy human macula. The number of IL-33+ Müller cells was increased in areas of retinal atrophy and in the vitreous of patients with AMD [67]. IL-33 is released from Müller cells in response to stress, which leads to CCL2 expression, macrophage recruitment, and tissue damage [68]. In cases of significant photoreceptor cell loss, reactive Müller glia can also insert their processes into the empty spaces left behind by lost photoreceptor cells to form a protective barrier around the remaining inner retina. Consequently, these barriers interfere with neuronal survival by limiting synaptic remodeling of the remaining neural retina ([64]; reviewed in [69]). In cases of transplantation, these barriers could significantly hinder the successful integration of donor cells. For example, high levels of reactive gliosis has been observed in response to transplantation of both fetal retina sheets [70] and photoreceptors [64], resulting in disrupted contact between donor and recipient tissue. Gliosis therefore represents a major hurdle to overcome when considering cell transplantation to treat various forms of retinal degeneration, as both an inflammatory environment and a physical barrier must be overcome in many cases.

There is increasing evidence that Müller glia–microglia crosstalk drives neuroinflammation [71, 72]. Recent studies in diabetic mice show that retinal inflammation is dependent upon CD40 activation in Müller glia [73, 74]. CD40 activation induces release of ATP by Müller glia leading to activation of P2X7 purinergic receptors on retinal microglia and subsequent inflammatory cytokine secretion. On the other hand, Müller glia cocultured with activated microglia increase expression of adhesion molecules (VCAM-1 and ICAM-1), inflammatory cytokines (CCL2 and CCL3), and proinflammatory factors such as IL-1β, IL-6, and iNOS [75].

Recently there has been a lot of focus on inflammasome activation in the CNS. It has been shown that paracrine factors secreted by reactive migrating subretinal microglia can trigger NLRP3 inflammasome activation in RPE cells [76, 77]. Proinflammatory factors such as TNF-α, IL-1α, and reactive oxygen species (ROS), secreted by reactive microglia, drive RPE cells to trigger inflammasome formation by activating the NFkB pathway. This promotes expression of NLRP3, and precursor forms of IL-1β and IL-18. Once the inflammasome is primed, it can be activated by various factors, including complement factors C1q [78], C3a [79], and the MAC [80], as well as amyloid β in drusens [81], and ROS and ATP from reactive microglia [62, 82]. Successful assembly of the inflammasome triggers activation of procaspase-1 into active caspase-1 and the cleavage of pro-IL-1β and pro-IL-18 into bioactive peptides [83]. Taken together, inflammasome activation is becoming more widely recognized as an important regulator of inflammation and tissue damage in the brain as well as the retina.

Another phenomenon that occurs with normal aging is the accumulation of the membrane attack complex (MAC), also known as the terminal complement complex (C5b-9), in the human choriocapillaris. Although the precise cause and direct repercussions of MAC accumulation in the eye have yet to be determined, the increase in complement activation over time may result in chronic low-grade inflammation, pathogenic angiogenesis, and significant tissue damage [84].

4.4   Cell Rejection Response Following Transplantation of Allogenic Cells in the Retina

Regardless of the source, inflammation often influences or is elevated by cell loss in many retinal degenerative diseases. While some diseases may benefit from immune modulation alone (discussed below), many retinal diseases progress to cell loss and require cell replacement to restore or improve vision. Cell rejection is a major concern when planning and/or executing cell replacement strategies; this is especially true when utilizing nonautologous donors. In such cases, the use of immune-suppressive strategies are often required.

For example, in mouse studies looking at photoreceptor integration, even partially matched donor cells were lost over 6–12 months post-transplantation with increasing inflammatory cell infiltration. However, the observed inflammation could be slowed by chemical immunosuppression using cyclosporine [85]. While integration efficiency may be overestimated in a number of transplantation studies due to the recent discovery of material transfer between transplanted cells and host retina[86–89], the commonly observed presence of inflammatory cells in the retina suggests a rejection response[85, 90]. Another study by Magdalena Seiler's group [91] demonstrated improved integration in their immunodeficient rat model following human cell transplantation. The group transplanted human neural progenitor cells into rhodopsin mutant Foxn1^{rnu} rats and found long-term survival and proliferation of the transplanted cells in the host subretinal space. Our lab looked to further assess human photoreceptor integration following transplantation in wild-type, IL2r$\gamma^{-/-}$, and Crx^{tvrm65} mice [92]. Following transplantation in IL2r$\gamma^{-/-}$ mice, which lack natural-killer (NK) and dendritic cells (DC) and have immature B and T cells [93], we observed robust integration of human cells, which was confirmed by the presence of human-specific markers [92, 94]. Transplantation in Crx^{tvrm65} mice bred onto the IL2r$\gamma^{-/-}$ background resulted in long-term donor cell survival, as far out as 9 months post-transplantation. This successful transplantation model allowed us to assess functional efficiency of the transplanted photoreceptors. We observed robust improvements in pupillary light responses and light-mediated CNS neuronal activity [92].

Several studies looking at cell rejection after RPE cell transplantation have been carried out in large animal models. For example, Masayo Takahashi's group compared allogeneic MHC-matched induced pluripotent stem cell (iPSC)-derived RPE transplants to mismatched RPE transplants in monkeys [95]. They observed that MHC-mismatched allografts did not survive following transplantation. Postmortem analysis revealed retinae containing inflammatory nodules consisting of inflammatory cells, including macrophages, inflammatory APCs, and CD3+ T cells, around the RPE grafts. In contrast, the transplanted retina allografts showed good survival with very few immune cells and APCs around the graft site. Further analysis by the same group showed RPE-specific alloantibodies in immune attack animal models that had B-cell (CD20+ and/or CD40+ cells) invasion in the retina at transplantation sites [96]. More recently, Trevor McGill's group similarly looked at allogeneic iPSC-derived RPE transplants injected into the subretinal space in non-immunosuppressed rhesus macaques [97]. They observed a strong inflammatory response localized to the site of transplanted cells within 4 days post-transplantation, which persisted for 2 months post-transplantation. This response was associated with the death of the transplanted RPE cells and the formation of subretinal plaques. Similar results have also been reported in mini-pigs [98]. In those studies, allogeneic transplantation of iPSC-derived RPE cells into mini-pigs resulted in an innate immune response, which was associated with the presence of activated macrophages and increased TGF-β levels in the vitreous of injected animals [98].

4.5   Strategies to Circumvent Cell Rejection

As described above, the immune-privilege status of the retina and RPE are relative and affected by factors that will likely require some form of intervention to promote survival of transplanted stem cell derivatives (Fig. 4.2). The most obvious approach, which is relatively difficult to implement in practice, is to use patient-specific autologous iPSC-derived cells [99]. While this method would provide the best host-recipient match, the procedure may be cost- and time-prohibitive, as generating the cells and testing for the presence of reprogramming-induced mutations can be time-consuming. Additionally, it is possible for protein expression changes or epigenetic changes to arise during the reprogramming process, and exposure to media components in culture could potentially generate immunogenic antigens on the surface of the cells [100]. As an alternative to the use of autologous cell lines, pools of HLA-matched lines could be generated. This strategy would require the identification of individuals who are homozygous for various HLA alleles and generate iPSC lines that would match large segments of the transplant recipient population [101, 102].

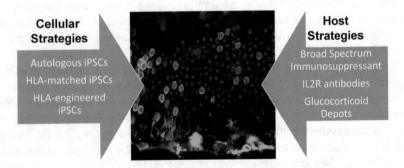

Fig. 4.2 Strategies to prevent cell rejection. Successful integration of transplanted photoreceptors will require a careful and focused approach to prevent cell rejection. These include approaches to reduce the immunogenicity of cells and also chemotherapeutic options for the hosts. The central image shows human GFP-expressing photoreceptors integrated into a mouse host retina 3-months post-transplantation

It has been estimated that about 55 donor iPSCs would match over 80% of HLA genotypes in the Japanese population [103], or a bank of 150 HLA-genotyped iPSCs could match 93% of the UK population based on HLA-A, B, and DR homozygous combinations [102]. Even for a diverse region like the state of California in the US, approximately 80 lines would cover almost 50% of the population [104]. Of course, the challenge here lies in identifying those homozygous individuals and collecting tissue or blood to generate these super-donor stem cell or retinal cell pools. A recently published study working toward this goal analyzed cord blood bank tissue to identify HLA matches within a Korean population [105]. After looking at over 4200 samples, the researchers identified the 10 most common homozygous HLA haplotypes and generated iPSCs from these samples. It was estimated that these 10 lines would match over 40% of the target Korean population [105]. An alternative approach being actively pursued is the generation of HLA-engineered lines, which involves knock-down of both alleles of *β2-microglobulin* [106] or *HLA-E* in iPSC lines [107]. This was tested in proof-of-principal studies using human iPSC-derived RPE cells that overexpressed HLA-E antigen. The found that the transplantations using the overexpressed line had much less natural-killer cells activation compared to controls [108]. This would circumvent the need to identify super-donors within the general population.

Alternatively, or in combination with providing HLA-matched cells, more effective immunosuppressive strategies should be considered. Ideally, immunosuppression efforts would need to prevent both acute and chronic rejection while maximizing allograft survival and long-term function. Based on our own work, IL-2 receptor antagonists, which are antibodies directed against the α subunit of the CD25 cell surface protein (i.e., IL-2 receptor), would be beneficial during transplantation, especially when utilizing direct delivery into the eye [92, 94]. A chimeric IL-2 receptor antibody, known as Basiliximab, is currently in use within the US [109–111]. Glucocorticoids may also play a critical role in maintaining immunosuppression and preventing chronic donor cell rejection[112–114]. As with Basiliximab, glucocorticoids would ideally be delivered locally within the eye, as is commonly carried out for the treatment of uveitis; direct delivery will help prevent systemic side-effects. One important consideration when using local glucocorticoids, however, is the increased risk of developing cataracts and ocular hypertension [115].

4.6   Overcoming Inflammation in the Retina

As described above, aging or chronic degenerative conditions lead to inflammatory environments that are detrimental to stem cell-based repairs. It goes to reason that (1) immune modulation aimed at harnessing the prorepair functions of innate immune cells or (2) generating anti-inflammatory environments will likely enhance regenerative efforts.

4.6.1 Immunomodulatory Factors

Work done by Neves et al. provides evidence for immune modulation as a key factor in tissue regeneration [116]. Specifically, this work reveals the importance of mesencephalic astrocyte-derived neurotrophic factor (MANF) as a regulator of the immune response during tissue repair and transplantation. MANF acts to promote the M2 prorepair activation state of innate immune cells in the eye (Fig. 4.3). Our study tested and found that MANF, through its immunomodulatory effects, delays retinal degeneration in mutliple differing models of photoreceptor loss in mice as well as a number of *Drosophila* models of retinal degeneration. Additionally, we found that use of MANF recombinant protein as an adjuvant promotes integration of transplanted photoreceptor cells in mice. Similarly, insulin-like growth factor 1 (IGF-1) has been shown to enhance the regenerative environment in skeletal muscles through immune modulation by alleviating inflammation and promoting anti-inflammatory phenotypes of macrophages [117]. Other factors, such as nerve growth factor [118], brain-derived neurotrophic factor [119], glial-derived ciliary neurotrophic factor [120], and leukemia inhibitory factor [121] have been shown to play a role in reducing inflammation and thereby may create a receptive environment for regenerative therapies.

4.6.2 Synthetic Immunomodulation

Minocyclin is a second-generation tetracycline antibiotic that exhibits anti-inflammatory properties by attenuating the expression of proinflammatory factors, including IL-6, IL-1β, and TLR-2 in murine brain [122–124]. Minocycline also inhibits the upregulation and release of IL-1β, TNF-α, and nitric oxide following bacterial LPS exposure in retinal microglia [125]. Additional studies have shown

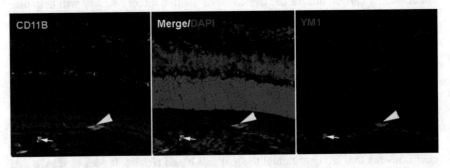

Fig. 4.3 MANF drives M2 activation of macrophages. Image of the mouse retina following human recombinant injection in a light-damaged animal. Most CD11B-expressing macrophages (green) in the subretinal space (arrowheads) and in the choroid (arrows) express M2 marker YM1 (red) following MANF exposure

that minocyclin (1) promotes photoreceptor survival following acute light stress by inhibiting proinflammatory activation of immune cells in the eye [126], (2) rescues retinal degeneration in juvenile neuronal ceroid lipofuscinosis mice [127], and (3) attenuates retinal degeneration in Abca4$^{-/-}$Rdh8$^{-/-}$ mice [128].

Similarly, translocator protein (18 kDa; TSPO) is a mitochondrial protein expressed on reactive glial cells [129]. A synthetic and highly specific TSPO ligand, XBD173 (AC-5216, emapunil), has been shown to attenuate microglial reactivity in the acute white light-induced retinal degeneration mouse model [130]. XBD173 treatment also reduced the expression of proinflammatory genes *CCL2* and *IL6* in LPS-challenged microglia in vitro [129].

Acting as a synthetic progesterone, Norgestrel exerts neuroprotection against retinal degeneration in an acute light-induced retinal degeneration mouse model and in the *rd*10 mouse model of retinitis pigmentosa [131, 132]. Norgestrel works directly on microglia, suppressing expression of proinflammatory cytokines, chemokines, and nitric oxide, thereby abrogating proinflammatory microglia-driven photoreceptor death [133] and increasing prosurvival factors including LIF [134] and bFGF [135].

Interferon beta (IFN-β) has been described to have potent immunomodulatory functions on microglia in the CNS. Mice lacking the *IFN-β* gene or its receptor display extensive microglia activation [136]. In contrast, IFN-β treatment strongly inhibits retinal microglial activation and enhances the transition to an M2-type, prorepair microglia phenotype in a mouse model of laser-induced choroidal neovascularization [137]. These effects have also been observed in larger animal models of choroidal neovascularization, such as rabbit and monkey [138, 139].

4.7 Conclusion

In this chapter, we have reviewed literature demonstrating the importance of immune modulation in promoting a receptive microenvironment for regenerative therapies. Understanding and overcoming these immune barriers will bring us another step closer to restoring vision in patients. Although each disease and cell type may require a slightly different approach, once the appropriate method has been established, reducing immune response should significantly enhance exogenous cell integration after transplantation within the retina and thereby enhance functional vision.

References

 1. Davis RJ, Blenkinsop TA, Campbell M, Borden SM, Charniga CJ, Lederman PL, Frye AM, Aguilar V, Zhao C, Naimark M et al (2016) Human RPE stem cell-derived RPE preserves photoreceptors in the Royal College of Surgeons rat: method for quantifying the area of photoreceptor sparing. J Ocul Pharmacol Ther 32:304–309

2. Khristov V, Maminishkis A, Amaral J, Rising A, Bharti K, Miller S (2018) Validation of iPS cell-derived RPE tissue in animal models. Adv Exp Med Biol 1074:633–640
3. Lamba DA, Karl MO, Ware CB, Reh TA (2006) Efficient generation of retinal progenitor cells from human embryonic stem cells. Proc Natl Acad Sci U S A 103:12769–12774
4. Lamba DA, Gust J, Reh TA (2009) Transplantation of human embryonic stem cell-derived photoreceptors restores some visual function in Crx-deficient mice. Cell Stem Cell 4:73–79
5. McGill TJ, Bohana-Kashtan O, Stoddard JW, Andrews MD, Pandit N, Rosenberg-Belmaker LR, Wiser O, Matzrafi L, Banin E, Reubinoff B et al (2017) Long-term efficacy of GMP grade xeno-free hESC-derived RPE cells following transplantation. Transl Vis Sci Technol 6:17
6. Meyer JS, Howden SE, Wallace KA, Verhoeven AD, Wright LS, Capowski EE, Pinilla I, Martin JM, Tian S, Stewart R et al (2011) Optic vesicle-like structures derived from human pluripotent stem cells facilitate a customized approach to retinal disease treatment. Stem Cells 29:1206–1218
7. Wu S, Chang K-C, Nahmou M, Goldberg JL (2018) Induced pluripotent stem cells promote retinal ganglion cell survival after transplant. Invest Ophthalmol Vis Sci 59:1571–1576
8. Collins KH, Herzog W, Reimer RA, Reno CR, Heard BJ, Hart DA (2018) Diet-induced obesity leads to pro-inflammatory alterations to the vitreous humour of the eye in a rat model. Inflamm Res 67:139–146
9. Niederkorn JY (2012) Ocular immune privilege and ocular melanoma: parallel universes or immunological plagiarism? Front Immunol 3:148
10. Osuský R, Walker SM, Ryan SJ (1996) Vitreous body affects activation and maturation of monocytes into macrophages. Graefes Arch Clin Exp Ophthalmol 234:637–642
11. Arroba AI, Campos-Caro A, Aguilar-Diosdado M, Valverde ÁM (2018) IGF-1, inflammation and retinal degeneration: a close network. Front Aging Neurosci 10:203
12. Sugita S, Futagami Y, Smith SB, Naggar H, Mochizuki M (2006) Retinal and ciliary body pigment epithelium suppress activation of T lymphocytes via transforming growth factor beta. Exp Eye Res 83:1459–1471
13. Sugita S, Horie S, Nakamura O, Futagami Y, Takase H, Keino H, Aburatani H, Katunuma N, Ishidoh K, Yamamoto Y et al (2008) Retinal pigment epithelium-derived CTLA-2alpha induces TGFbeta-producing T regulatory cells. J Immunol 1950(181):7525–7536
14. Zhou R, Horai R, Silver PB, Mattapallil MJ, Zárate-Bladés CR, Chong WP, Chen J, Rigden RC, Villasmil R, Caspi RR (2012) The living eye "disarms" uncommitted autoreactive T cells by converting them to FoxP3+ regulatory cells following local antigen recognition. J Immunol 1950(188):1742–1750
15. Forbes SJ, Rosenthal N (2014) Preparing the ground for tissue regeneration: from mechanism to therapy. Nat Med 20:857–869
16. Gordon S, Taylor PR (2005) Monocyte and macrophage heterogeneity. Nat Rev Immunol 5:953–964
17. Hanisch U-K, Kettenmann H (2007) Microglia: active sensor and versatile effector cells in the normal and pathologic brain. Nat Neurosci 10:1387–1394
18. Mosser DM, Edwards JP (2008) Exploring the full spectrum of macrophage activation. Nat Rev Immunol 8:958–969
19. London A, Cohen M, Schwartz M (2013) Microglia and monocyte-derived macrophages: functionally distinct populations that act in concert in CNS plasticity and repair. Front Cell Neurosci 7:34
20. Gordon S, Martinez FO (2010) Alternative activation of macrophages: mechanism and functions. Immunity 32:593–604
21. Martinez FO, Helming L, Gordon S (2009) Alternative activation of macrophages: an immunologic functional perspective. Annu Rev Immunol 27:451–483
22. Chen M, Luo C, Zhao J, Devarajan G, Xu H (2019) Immune regulation in the aging retina. Prog Retin Eye Res 69:159–172
23. Li L, Eter N, Heiduschka P (2015) The microglia in healthy and diseased retina. Exp Eye Res 136:116–130

24. Rathnasamy G, Foulds WS, Ling E-A, Kaur C (2018) Retinal microglia—a key player in healthy and diseased retina. Prog Neurobiol 173:18–40
25. Hagemeyer N, Hanft K-M, Akriditou M-A, Unger N, Park ES, Stanley ER, Staszewski O, Dimou L, Prinz M (2017) Microglia contribute to normal myelinogenesis and to oligodendrocyte progenitor maintenance during adulthood. Acta Neuropathol (Berl) 134:441–458
26. Kim S-Y (2015) Retinal phagocytes in age-related macular degeneration. Macrophage 2:e698
27. Prinz M, Erny D, Hagemeyer N (2017) Ontogeny and homeostasis of CNS myeloid cells. Nat Immunol 18:385–392
28. Sedel F, Béchade C, Vyas S, Triller A (2004) Macrophage-derived tumor necrosis factor alpha, an early developmental signal for motoneuron death. J Neurosci 24:2236–2246
29. Casano AM, Albert M, Peri F (2016) Developmental apoptosis mediates entry and positioning of microglia in the zebrafish brain. Cell Rep 16:897–906
30. Silverman SM, Wong WT (2018) Microglia in the retina: roles in development, maturity, and disease. Annu Rev Vis Sci 4:45–77
31. Wang X, Zhao L, Zhang J, Fariss RN, Ma W, Kretschmer F, Wang M, Qian HH, Badea TC, Diamond JS et al (2016) Requirement for microglia for the maintenance of synaptic function and integrity in the mature retina. J Neurosci 36:2827–2842
32. Xu J, Wang T, Wu Y, Jin W, Wen Z (2016) Microglia colonization of developing zebrafish midbrain is promoted by apoptotic neuron and lysophosphatidylcholine. Dev Cell 38:214–222
33. Arnold T, Betsholtz C (2013) The importance of microglia in the development of the vasculature in the central nervous system. Vasc Cell 5:4
34. Frost JL, Schafer DP (2016) Microglia: architects of the developing nervous system. Trends Cell Biol 26:587–597
35. Nakanishi M, Niidome T, Matsuda S, Akaike A, Kihara T, Sugimoto H (2007) Microglia-derived interleukin-6 and leukaemia inhibitory factor promote astrocytic differentiation of neural stem/progenitor cells. Eur J Neurosci 25:649–658
36. Ma W, Zhang Y, Gao C, Fariss RN, Tam J, Wong WT (2017) Monocyte infiltration and proliferation reestablish myeloid cell homeostasis in the mouse retina following retinal pigment epithelial cell injury. Sci Rep 7:8433
37. O'Koren EG, Mathew R, Saban DR (2016) Fate mapping reveals that microglia and recruited monocyte-derived macrophages are definitively distinguishable by phenotype in the retina. Sci Rep 6:20636
38. MacDonald RB, Charlton-Perkins M, Harris WA (2017) Mechanisms of Müller glial cell morphogenesis. Curr Opin Neurobiol 47:31–37
39. Subirada PV, Paz MC, Ridano ME, Lorenc VE, Vaglienti MV, Barcelona PF, Luna JD, Sánchez MC (2018) A journey into the retina: Müller glia commanding survival and death. Eur J Neurosci 47:1429–1443
40. Wang J, O'Sullivan ML, Mukherjee D, Puñal VM, Farsiu S, Kay JN (2017) Anatomy and spatial organization of Müller glia in mouse retina. J Comp Neurol 525:1759–1777
41. Wing K, Sakaguchi S (2010) Regulatory T cells exert checks and balances on self tolerance and autoimmunity. Nat Immunol 11:7–13
42. Josefowicz SZ, Lu L-F, Rudensky AY (2012) Regulatory T cells: mechanisms of differentiation and function. Annu Rev Immunol 30:531–564
43. Burzyn D, Kuswanto W, Kolodin D, Shadrach JL, Cerletti M, Jang Y, Sefik E, Tan TG, Wagers AJ, Benoist C et al (2013) A special population of regulatory T cells potentiates muscle repair. Cell 155:1282–1295
44. Burzyn D, Benoist C, Mathis D (2013) Regulatory T cells in nonlymphoid tissues. Nat Immunol 14:1007–1013
45. Holers VM (2014) Complement and its receptors: new insights into human disease. Annu Rev Immunol 32:433–459
46. Haynes T, Luz-Madrigal A, Reis ES, Echeverri Ruiz NP, Grajales-Esquivel E, Tzekou A, Tsonis PA, Lambris JD, Del Rio-Tsonis K (2013) Complement anaphylatoxin C3a is a potent inducer of embryonic chick retina regeneration. Nat Commun 4:2312

47. Hawksworth OA, Coulthard LG, Woodruff TM (2017) Complement in the fundamental processes of the cell. Mol Immunol 84:17–25
48. Mullins RF, Schoo DP, Sohn EH, Flamme-Wiese MJ, Workamelahu G, Johnston RM, Wang K, Tucker BA, Stone EM (2014) The membrane attack complex in aging human choriocapillaris: relationship to macular degeneration and choroidal thinning. Am J Pathol 184:3142–3153
49. Karlstetter M, Scholz R, Rutar M, Wong WT, Provis JM, Langmann T (2015) Retinal microglia: just bystander or target for therapy? Prog Retin Eye Res 45:30–57
50. Hu R, Kagele DA, Huffaker TB, Runtsch MC, Alexander M, Liu J, Bake E, Su W, Williams MA, Rao DS et al (2014) miR-155 promotes T follicular helper cell accumulation during chronic, low-grade inflammation. Immunity 41:605–619
51. Cuenca N, Fernández-Sánchez L, Campello L, Maneu V, De la Villa P, Lax P, Pinilla I (2014) Cellular responses following retinal injuries and therapeutic approaches for neurodegenerative diseases. Prog Retin Eye Res 43:17–75
52. Bringmann A, Pannicke T, Grosche J, Francke M, Wiedemann P, Skatchkov SN, Osborne NN, Reichenbach A (2006) Müller cells in the healthy and diseased retina. Prog Retin Eye Res 25:397–424
53. Martinez FO, Gordon S (2014) The M1 and M2 paradigm of macrophage activation: time for reassessment. F1000Prime Rep 6:13
54. Yin Y, Henzl MT, Lorber B, Nakazawa T, Thomas TT, Jiang F, Langer R, Benowitz LI (2006) Oncomodulin is a macrophage-derived signal for axon regeneration in retinal ganglion cells. Nat Neurosci 9:843–852
55. Suh H-S, Zhao M-L, Derico L, Choi N, Lee SC (2013) Insulin-like growth factor 1 and 2 (IGF1, IGF2) expression in human microglia: differential regulation by inflammatory mediators. J Neuroinflammation 10:37
56. Krause TA, Alex AF, Engel DR, Kurts C, Eter N (2014) VEGF-production by CCR2-dependent macrophages contributes to laser-induced choroidal neovascularization. PLoS One 9:e94313
57. Liu J, Copland DA, Horie S, Wu W-K, Chen M, Xu Y, Paul Morgan B, Mack M, Xu H, Nicholson LB et al (2013) Myeloid cells expressing VEGF and arginase-1 following uptake of damaged retinal pigment epithelium suggests potential mechanism that drives the onset of choroidal angiogenesis in mice. PLoS One 8:e72935
58. Miron VE, Boyd A, Zhao J-W, Yuen TJ, Ruckh JM, Shadrach JL, van Wijngaarden P, Wagers AJ, Williams A, Franklin RJM et al (2013) M2 microglia and macrophages drive oligodendrocyte differentiation during CNS remyelination. Nat Neurosci 16:1211–1218
59. Sica A, Mantovani A (2012) Macrophage plasticity and polarization: in vivo veritas. J Clin Invest 122:787–795
60. Cherry JD, Olschowka JA, O'Banion MK (2014) Neuroinflammation and M2 microglia: the good, the bad, and the inflamed. J Neuroinflammation 11:98
61. Shaw AC, Goldstein DR, Montgomery RR (2013) Age-dependent dysregulation of innate immunity. Nat Rev Immunol 13:875–887
62. Madeira MH, Rashid K, Ambrósio AF, Santiago AR, Langmann T (2018) Blockade of microglial adenosine A2A receptor impacts inflammatory mechanisms, reduces ARPE-19 cell dysfunction and prevents photoreceptor loss in vitro. Sci Rep 8:2272
63. Song D, Sulewski ME, Wang C, Song J, Bhuyan R, Sterling J, Clark E, Song W-C, Dunaief JL (2017) Complement C5a receptor knockout has diminished light-induced microglia/macrophage retinal migration. Mol Vis 23:210–218
64. Hippert C, Graca AB, Barber AC, West EL, Smith AJ, Ali RR, Pearson RA (2015) Müller glia activation in response to inherited retinal degeneration is highly varied and disease-specific. PLoS One 10:e0120415
65. Crespo D, Asher RA, Lin R, Rhodes KE, Fawcett JW (2007) How does chondroitinase promote functional recovery in the damaged CNS? Exp Neurol 206:159–171
66. McKeon RJ, Schreiber RC, Rudge JS, Silver J (1991) Reduction of neurite outgrowth in a model of glial scarring following CNS injury is correlated with the expression of inhibitory molecules on reactive astrocytes. J Neurosci 11:3398–3411

67. Xi H, Katschke KJ, Li Y, Truong T, Lee WP, Diehl L, Rangell L, Tao J, Arceo R, Eastham-Anderson J et al (2016) IL-33 amplifies an innate immune response in the degenerating retina. J Exp Med 213:189–207

68. Rutar M, Natoli R, Provis JM (2012) Small interfering RNA-mediated suppression of Ccl2 in Müller cells attenuates microglial recruitment and photoreceptor death following retinal degeneration. J Neuroinflammation 9:221

69. Graca AB, Hippert C, Pearson RA (2018) Müller Glia reactivity and development of gliosis in response to pathological conditions. Adv Exp Med Biol 1074:303–308

70. Radtke ND, Aramant RB, Seiler M, Petry HM (1999) Preliminary report: indications of improved visual function after retinal sheet transplantation in retinitis pigmentosa patients. Am J Ophthalmol 128:384–387

71. Arroba AI, Alvarez-Lindo N, van Rooijen N, de la Rosa EJ (2014) Microglia-Müller glia crosstalk in the rd10 mouse model of retinitis pigmentosa. Adv Exp Med Biol 801:373–379

72. Roche SL, Ruiz-Lopez AM, Moloney JN, Byrne AM, Cotter TG (2018) Microglial-induced Müller cell gliosis is attenuated by progesterone in a mouse model of retinitis pigmentosa. Glia 66:295–310

73. Portillo J-AC, Lopez Corcino Y, Miao Y, Tang J, Sheibani N, Kern TS, Dubyak GR, Subauste CS (2017) CD40 in retinal Müller cells induces P2X7-dependent cytokine expression in macrophages/microglia in diabetic mice and development of early experimental diabetic retinopathy. Diabetes 66:483–493

74. Samuels IS, Portillo J-AC, Miao Y, Kern TS, Subauste CS (2017) Loss of CD40 attenuates experimental diabetes-induced retinal inflammation but does not protect mice from electroretinogram defects. Vis Neurosci 34:E009

75. Wang J, Westenskow PD, Fang M, Friedlander M, Siuzdak G (2016) Quantitative metabolomics of photoreceptor degeneration and the effects of stem cell-derived retinal pigment epithelium transplantation. Philos Trans Math Phys Eng Sci 374:20150376

76. Kerur N, Hirano Y, Tarallo V, Fowler BJ, Bastos-Carvalho A, Yasuma T, Yasuma R, Kim Y, Hinton DR, Kirschning CJ et al (2013) TLR-independent and P2X7-dependent signaling mediate Alu RNA-induced NLRP3 inflammasome activation in geographic atrophy. Invest Ophthalmol Vis Sci 54:7395–7401

77. Nebel C, Aslanidis A, Rashid K, Langmann T (2017) Activated microglia trigger inflammasome activation and lysosomal destabilization in human RPE cells. Biochem Biophys Res Commun 484:681–686

78. Doyle SL, Campbell M, Ozaki E, Salomon RG, Mori A, Kenna PF, Farrar GJ, Kiang A-S, Humphries MM, Lavelle EC et al (2012) NLRP3 has a protective role in age-related macular degeneration through the induction of IL-18 by drusen components. Nat Med 18:791–798

79. Asgari E, Le Friec G, Yamamoto H, Perucha E, Sacks SS, Köhl J, Cook HT, Kemper C (2013) C3a modulates IL-1β secretion in human monocytes by regulating ATP efflux and subsequent NLRP3 inflammasome activation. Blood 122:3473–3481

80. Triantafilou K, Hughes TR, Triantafilou M, Morgan BP (2013) The complement membrane attack complex triggers intracellular Ca^{2+} fluxes leading to NLRP3 inflammasome activation. J Cell Sci 126:2903–2913

81. Halle A, Hornung V, Petzold GC, Stewart CR, Monks BG, Reinheckel T, Fitzgerald KA, Latz E, Moore KJ, Golenbock DT (2008) The NALP3 inflammasome is involved in the innate immune response to amyloid-beta. Nat Immunol 9:857–865

82. Heid ME, Keyel PA, Kamga C, Shiva S, Watkins SC, Salter RD (2013) Mitochondrial reactive oxygen species induces NLRP3-dependent lysosomal damage and inflammasome activation. J Immunol 191:5230–5238

83. Guo H, Callaway JB, Ting JP-Y (2015) Inflammasomes: mechanism of action, role in disease, and therapeutics. Nat Med 21:677–687

84. Zeng S, Whitmore SS, Sohn EH, Riker MJ, Wiley LA, Scheetz TE, Stone EM, Tucker BA, Mullins RF (2016) Molecular response of chorioretinal endothelial cells to complement injury: implications for macular degeneration. J Pathol 238:446–456

85. West EL, Pearson RA, Barker SE, Luhmann UFO, Maclaren RE, Barber AC, Duran Y, Smith AJ, Sowden JC, Ali RR (2010) Long-term survival of photoreceptors transplanted into the adult murine neural retina requires immune modulation. Stem Cells 28:1997–2007

86. Pearson RA, Gonzalez-Cordero A, West EL, Ribeiro JR, Aghaizu N, Goh D, Sampson RD, Georgiadis A, Waldron PV, Duran Y et al (2016) Donor and host photoreceptors engage in material transfer following transplantation of post-mitotic photoreceptor precursors. Nat Commun 7:13029

87. Santos-Ferreira T, Llonch S, Borsch O, Postel K, Haas J, Ader M (2016) Retinal transplantation of photoreceptors results in donor-host cytoplasmic exchange. Nat Commun 7:13028

88. Singh MS, Balmer J, Barnard AR, Aslam SA, Moralli D, Green CM, Barnea-Cramer A, Duncan I, MacLaren RE (2016) Transplanted photoreceptor precursors transfer proteins to host photoreceptors by a mechanism of cytoplasmic fusion. Nat Commun 7:13537

89. Waldron PV, Di Marco F, Kruczek K, Ribeiro J, Graca AB, Hippert C, Aghaizu ND, Kalargyrou AA, Barber AC, Grimaldi G et al (2018) Transplanted donor- or stem cell-derived cone photoreceptors can both integrate and undergo material transfer in an environment-dependent manner. Stem Cell Rep 10:406–421

90. Jiang LQ, Jorquera M, Streilein JW, Ishioka M (1995) Unconventional rejection of neural retinal allografts implanted into the immunologically privileged site of the eye. Transplantation 59:1201–1207

91. Seiler MJ, Aramant RB, Jones MK, Ferguson DL, Bryda EC, Keirstead HS (2014) A new immunodeficient pigmented retinal degenerate rat strain to study transplantation of human cells without immunosuppression. Graefes Arch Clin Exp Ophthalmol 252:1079–1092

92. Zhu J, Cifuentes H, Reynolds J, Lamba DA (2017) Immunosuppression via loss of IL2rgamma enhances long-term functional integration of hESC-derived photoreceptors in the mouse retina. Cell Stem Cell 20:374–384.e5

93. Cao X, Shores EW, Hu-Li J, Anver MR, Kelsall BL, Russell SM, Drago J, Noguchi M, Grinberg A, Bloom ET et al (1995) Defective lymphoid development in mice lacking expression of the common cytokine receptor gamma chain. Immunity 2:223–238

94. Zhu J, Reynolds J, Garcia T, Cifuentes H, Chew S, Zeng X, Lamba DA (2018) Generation of transplantable retinal photoreceptors from a current good manufacturing practice-manufactured human induced pluripotent stem cell line. Stem Cells Transl Med 7:210–219

95. Sugita S, Iwasaki Y, Makabe K, Kamao H, Mandai M, Shiina T, Ogasawara K, Hirami Y, Kurimoto Y, Takahashi M (2016) Successful transplantation of retinal pigment epithelial cells from MHC homozygote iPSCs in MHC-matched models. Stem Cell Rep 7:635–648

96. Sugita S, Makabe K, Fujii S, Iwasaki Y, Kamao H, Shiina T, Ogasawara K, Takahashi M (2017) Detection of retinal pigment epithelium-specific antibody in iPSC-derived retinal pigment epithelium transplantation models. Stem Cell Rep 9:1501–1515

97. McGill TJ, Stoddard J, Renner LM, Messaoudi I, Bharti K, Mitalipov S, Lauer A, Wilson DJ, Neuringer M (2018) Allogeneic iPSC-derived RPE cell graft failure following transplantation into the subretinal space in nonhuman primates. Invest Ophthalmol Vis Sci 59:1374–1383

98. Sohn EH, Jiao C, Kaalberg E, Cranston C, Mullins RF, Stone EM, Tucker BA (2015) Allogenic iPSC-derived RPE cell transplants induce immune response in pigs: a pilot study. Sci Rep 5:11791

99. Mandai M, Watanabe A, Kurimoto Y, Hirami Y, Morinaga C, Daimon T, Fujihara M, Akimaru H, Sakai N, Shibata Y et al (2017) Autologous induced stem-cell-derived retinal cells for macular degeneration. N Engl J Med 376:1038–1046

100. Ruiz S, Diep D, Gore A, Panopoulos AD, Montserrat N, Plongthongkum N, Kumar S, Fung H-L, Giorgetti A, Bilic J et al (2012) Identification of a specific reprogramming-associated epigenetic signature in human induced pluripotent stem cells. Proc Natl Acad Sci U S A 109:16196–16201

101. Gourraud P-A, Gilson L, Girard M, Peschanski M (2012) The role of human leukocyte antigen matching in the development of multiethnic "haplobank" of induced pluripotent stem cell lines. Stem Cells Dayt Ohio 30:180–186

102. Turner M, Leslie S, Martin NG, Peschanski M, Rao M, Taylor CJ, Trounson A, Turner D, Yamanaka S, Wilmut I (2013) Toward the development of a global induced pluripotent stem cell library. Cell Stem Cell 13:382–384

103. Nakajima F, Tokunaga K, Nakatsuji N (2007) Human leukocyte antigen matching estimations in a hypothetical bank of human embryonic stem cell lines in the Japanese population for use in cell transplantation therapy. Stem Cells Dayt Ohio 25:983–985

104. Pappas DJ, Gourraud P-A, Gall CL, Laurent J, Trounson A, DeWitt N, Talib S (2015) Proceedings: Human leukocyte antigen haplo-homozygous induced pluripotent stem cell haplobank modeled after the California population: evaluating matching in a multiethnic and admixed population. Stem Cells Transl Med 4:413–418

105. Lee S, Huh JY, Turner DM, Lee S, Robinson J, Stein JE, Shim SH, Hong CP, Kang MS, Nakagawa M et al (2018) Repurposing the cord blood bank for haplobanking of HLA-homozygous iPSCs and their usefulness to multiple populations. Stem Cells 36:1552–1566

106. Riolobos L, Hirata RK, Turtle CJ, Wang P-R, Gornalusse GG, Zavajlevski M, Riddell SR, Russell DW (2013) HLA engineering of human pluripotent stem cells. Mol Ther 21:1232–1241

107. Gornalusse GG, Hirata RK, Funk SE, Riolobos L, Lopes VS, Manske G, Prunkard D, Colunga AG, Hanafi L-A, Clegg DO et al (2017) HLA-E-expressing pluripotent stem cells escape allogeneic responses and lysis by NK cells. Nat Biotechnol 35:765–772

108. Sugita S, Makabe K, Iwasaki Y, Fujii S, Takahashi M (2018) Natural killer cell inhibition by HLA-E molecules on induced pluripotent stem cell-derived retinal pigment epithelial cells. Invest Ophthalmol Vis Sci 59:1719–1731

109. de Ataide EC, Perales SR, Bortoto JB, Peres MAO, Filho FC, Stucchi RSB, Udo E, Boin IFSF (2017) Immunomodulation, acute renal failure, and complications of basiliximab use after liver transplantation: analysis of 114 patients and literature review. Transplant Proc 49:852–857

110. Kittipibul V, Tantrachoti P, Ongcharit P, Ariyachaipanich A, Siwamogsatham S, Sritangsirikul S, Thammanatsakul K, Puwanant S (2017) Low-dose basiliximab induction therapy in heart transplantation. Clin Transplant 31. https://doi.org/10.1111/ctr.13132

111. Zhang G-Q, Zhang C-S, Sun N, Lv W, Chen B-M, Zhang J-L (2017) Basiliximab application on liver recipients: a meta-analysis of randomized controlled trials. Hepatobiliary Pancreat Dis Int 16:139–146

112. Franchimont D (2004) Overview of the actions of glucocorticoids on the immune response: a good model to characterize new pathways of immunosuppression for new treatment strategies. Ann N Y Acad Sci 1024:124–137

113. Pan Q, Xu Q, Boylan NJ, Lamb NW, Emmert DG, Yang J-C, Tang L, Heflin T, Alwadani S, Eberhart CG et al (2015) Corticosteroid-loaded biodegradable nanoparticles for prevention of corneal allograft rejection in rats. J Control Release 201:32–40

114. Young AL, Rao SK, Cheng LL, Wong AKK, Leung ATS, Lam DSC (2002) Combined intravenous pulse methylprednisolone and oral cyclosporine A in the treatment of corneal graft rejection: 5-year experience. Eye (Lond) 16:304–308

115. Sen HN, Abreu FM, Louis TA, Sugar EA, Altaweel MM, Elner SG, Holbrook JT, Jabs DA, Kim RY, Kempen JH (2016) Cataract surgery outcomes in uveitis: the multicenter uveitis steroid treatment trial. Ophthalmology 123:183–190

116. Neves J, Zhu J, Sousa-Victor P, Konjikusic M, Riley R, Chew S, Qi Y, Jasper H, Lamba DA (2016) Immune modulation by MANF promotes tissue repair and regenerative success in the retina. Science 353:aaf3646

117. Tonkin J, Temmerman L, Sampson RD, Gallego-Colon E, Barberi L, Bilbao D, Schneider MD, Musarò A, Rosenthal N (2015) Monocyte/macrophage-derived IGF-1 orchestrates murine skeletal muscle regeneration and modulates autocrine polarization. Mol Ther 23:1189–1200

118. Prencipe G, Minnone G, Strippoli R, Pasquale LD, Petrini S, Caiello I, Manni L, Benedetti FD, Bracci-Laudiero L (2014) Nerve growth factor downregulates inflammatory response in human monocytes through TrkA. J Immunol 192:3345–3354

119. Calabrese F, Rossetti AC, Racagni G, Gass P, Riva MA, Molteni R (2014) Brain-derived neurotrophic factor: a bridge between inflammation and neuroplasticity. Front Cell Neurosci 8:430

120. Linker RA, Mäurer M, Gaupp S, Martini R, Holtmann B, Giess R, Rieckmann P, Lassmann H, Toyka KV, Sendtner M et al (2002) CNTF is a major protective factor in demyelinating CNS disease: a neurotrophic cytokine as modulator in neuroinflammation. Nat Med 8:620–624

121. Banner LR, Patterson PH, Allchorne A, Poole S, Woolf CJ (1998) Leukemia inhibitory factor is an anti-inflammatory and analgesic cytokine. J Neurosci 18:5456–5462

122. Henry CJ, Huang Y, Wynne A, Hanke M, Himler J, Bailey MT, Sheridan JF, Godbout JP (2008) Minocycline attenuates lipopolysaccharide (LPS)-induced neuroinflammation, sickness behavior, and anhedonia. J Neuroinflammation 5:15

123. Nikodemova M, Duncan ID, Watters JJ (2006) Minocycline exerts inhibitory effects on multiple mitogen-activated protein kinases and IkappaBalpha degradation in a stimulus-specific manner in microglia. J Neurochem 96:314–323

124. Yoon S-Y, Patel D, Dougherty PM (2012) Minocycline blocks lipopolysaccharide induced hyperalgesia by suppression of microglia but not astrocytes. Neuroscience 221:214–224

125. Wang AL, Yu ACH, Lau LT, Lee C, Wu LM, Zhu X, Tso MOM (2005) Minocycline inhibits LPS-induced retinal microglia activation. Neurochem Int 47:152–158

126. Scholz R, Sobotka M, Caramoy A, Stempfl T, Moehle C, Langmann T (2015) Minocycline counter-regulates pro-inflammatory microglia responses in the retina and protects from degeneration. J Neuroinflammation 12:209

127. Dannhausen K, Möhle C, Langmann T (2018) Immunomodulation with minocycline rescues retinal degeneration in juvenile Neuronal Ceroid Lipofuscinosis (jNCL) mice highly susceptible to light damage. Dis Model Mech 11:dmm.033597

128. Kohno H, Chen Y, Kevany BM, Pearlman E, Miyagi M, Maeda T, Palczewski K, Maeda A (2013) Photoreceptor proteins initiate microglial activation via Toll-like receptor 4 in retinal degeneration mediated by all-trans-retinal. J Biol Chem 288(21):15326–15341

129. Karlstetter M, Nothdurfter C, Aslanidis A, Moeller K, Horn F, Scholz R, Neumann H, Weber BHF, Rupprecht R, Langmann T (2014) Translocator protein (18 kDa) (TSPO) is expressed in reactive retinal microglia and modulates microglial inflammation and phagocytosis. J Neuroinflammation 11:3

130. Scholz R, Caramoy A, Bhuckory MB, Rashid K, Chen M, Xu H, Grimm C, Langmann T (2015) Targeting translocator protein (18 kDa) (TSPO) dampens pro-inflammatory microglia reactivity in the retina and protects from degeneration. J Neuroinflammation 12:201

131. Doonan F, O'Driscoll C, Kenna P, Cotter TG (2011) Enhancing survival of photoreceptor cells in vivo using the synthetic progestin Norgestrel. J Neurochem 118:915–927

132. Jackson ACW, Roche SL, Byrne AM, Ruiz-Lopez AM, Cotter TG (2016) Progesterone receptor signalling in retinal photoreceptor neuroprotection. J Neurochem 136:63–77

133. Roche SL, Wyse-Jackson AC, Gómez-Vicente V, Lax P, Ruiz-Lopez AM, Byrne AM, Cuenca N, Cotter TG (2016) Progesterone attenuates microglial-driven retinal degeneration and stimulates protective fractalkine-CX3CR1 signaling. PLoS One 11:e0165197

134. Byrne AM, Roche SL, Ruiz-Lopez AM, Jackson ACW, Cotter TG (2016) The synthetic progestin norgestrel acts to increase LIF levels in the rd10 mouse model of retinitis pigmentosa. Mol Vis 22:264–274

135. Wyse Jackson AC, Cotter TG (2016) The synthetic progesterone Norgestrel is neuroprotective in stressed photoreceptor-like cells and retinal explants, mediating its effects via basic fibroblast growth factor, protein kinase A and glycogen synthase kinase 3β signalling. Eur J Neurosci 43:899–911

136. Prinz M, Schmidt H, Mildner A, Knobeloch K-P, Hanisch U-K, Raasch J, Merkler D, Detje C, Gutcher I, Mages J et al (2008) Distinct and nonredundant in vivo functions of IFNAR on myeloid cells limit autoimmunity in the central nervous system. Immunity 28:675–686

137. Lückoff A, Caramoy A, Scholz R, Prinz M, Kalinke U, Langmann T (2016) Interferon-beta signaling in retinal mononuclear phagocytes attenuates pathological neovascularization. EMBO Mol Med 8:670–678
138. Kimoto T, Takahashi K, Tobe T, Fujimoto K, Uyama M, Sone S (2002) Effects of local administration of interferon-beta on proliferation of retinal pigment epithelium in rabbit after laser photocoagulation. Jpn J Ophthalmol 46:160–169
139. Tobe T, Takahashi K, Ohkuma H, Uyama M (1995) The effect of interferon-beta on experimental choroidal neovascularization. Nippon Ganka Gakkai Zasshi 99:571–581

Chapter 5
Advances in the Differentiation of Retinal Ganglion Cells from Human Pluripotent Stem Cells

Sarah K. Ohlemacher, Kirstin B. Langer, Clarisse M. Fligor, Elyse M. Feder, Michael C. Edler, and Jason S. Meyer

Abstract Human pluripotent stem cell (hPSC) technology has revolutionized the field of biology through the unprecedented ability to study the differentiation of human cells in vitro. In the past decade, hPSCs have been applied to study development, model disease, develop drugs, and devise cell replacement therapies for numerous biological systems. Of particular interest is the application of this technology to study and treat optic neuropathies such as glaucoma. Retinal ganglion cells (RGCs) are the primary cell type affected in these diseases, and once lost, they are unable to regenerate in adulthood. This necessitates the development of strategies to study the mechanisms of degeneration as well as develop translational therapeutic approaches to treat early- and late-stage disease progression. Numerous protocols have been established to derive RGCs from hPSCs, with the ability to generate large populations of human RGCs for translational applications. In this review, the key applications of hPSCs within the retinal field are described, including the use of these cells as developmental models, disease models, drug development, and finally, cell replacement therapies. In greater detail, the current report focuses on the differentiation of hPSC-derived RGCs and the many unique

S. K. Ohlemacher · K. B. Langer · C. M. Fligor · E. M. Feder
Department of Biology, Indiana University Purdue University Indianapolis,
Indianapolis, IN, USA

M. C. Edler
Department of Biology, Indiana University Purdue University Indianapolis,
Indianapolis, IN, USA

Department of Medical and Molecular Genetics, Indiana University, Indianapolis, IN, USA

J. S. Meyer (✉)
Department of Biology, Indiana University Purdue University Indianapolis,
Indianapolis, IN, USA

Department of Medical and Molecular Genetics, Indiana University, Indianapolis, IN, USA

Stark Neurosciences Research Institute, Indiana University, Indianapolis, IN, USA
e-mail: meyerjas@iupui.edu

© Springer Nature Switzerland AG 2019
K. Bharti (ed.), *Pluripotent Stem Cells in Eye Disease Therapy*,
Advances in Experimental Medicine and Biology 1186,
https://doi.org/10.1007/978-3-030-28471-8_5

characteristics associated with these cells in vitro including their genetic identifiers, their electrophysiological activity, and their morphological maturation. Also described is the current progress in the use of patient-specific hPSCs to study optic neuropathies affecting RGCs, with emphasis on the use of these RGCs for studying disease mechanisms and pathogenesis, drug screening, and cell replacement therapies in future studies.

Keywords hPSCs · Retinal ganglion cells · Pluripotent stem cells · Retina · Optic neuropathies

Retinal ganglion cells (RGCs) play a crucial role in transmitting visual information from the eye to the brain. This transduction pathway can be severed due to disease or injury, which can inhibit light information from reaching the appropriate processing centers, and further result in loss of vision and blindness. Damage to the RGCs can occur in response to injury to the tissue, as well as following the onset of diseases known as optic neuropathies. Such debilitating conditions lead to the degeneration and eventual loss of RGCs, as these cells do not possess the capacity to regenerate in adulthood.

To date, no therapies exist to delay or halt the progression of RGC degeneration. Furthermore, by the time a clinical diagnosis has been delivered to a patient, a significant percentage of the RGC population has already been irreversibly lost [1]. This shortcoming necessitates the development of strategies to study the progression of RGC degeneration and pathogenesis as well as develop translational therapeutic approaches targeting RGCs. Human pluripotent stem cells (hPSCs) serve as an attractive model for such studies as they can be derived from patient somatic sources and can provide an unlimited source of cells that can be differentiated to any cell type of the body [2, 3]. As such, the utilization of hPSCs as a model system has revolutionized the field of developmental biology, translational disease modeling, and personalized medicine [4–7].

5.1   Applications of hPSCs

hPSCs can be used as an impactful and resourceful developmental model as they allow access to some of the earliest time points of embryonic development that would otherwise be unavailable. Before the discovery of hPSC technology, our understanding of retinal development was largely informed by animal models, with a limited option for studying the retina in humans through the use of human fetal or postmortem tissue. However, obtaining such samples was associated with numerous difficulties, as they were only accessible at limited developmental time points and ethical and legal issues limited their availability. Following the discovery of hPSCs, studies have effectively demonstrated their use as a novel model to study the major stages of human retinogenesis, including the primitive eye field giving rise to the evaginating optic vesicle, as well as the development of an optic cup-like structure deemed retinal organoids [8–11]. These hPSCs give rise to distinct populations of retinal neurons which not only follow the temporal sequence of embryonic retinal development, but also recapitulate the cellular mosaicism and lamination of the in vivo retina, allowing for a more bonafide model to study retinal development and disease [9, 11–16]. Furthermore, patient-specific hPSC-derived retinal neurons can be used for studying cell-specific mechanisms and have future implications for studying regeneration of retinal tissue following injury or disease [14, 17–20].

The studies of optic neuropathies caused by genetic determinants using hPSCs are of particular interest as they are the result of known mutations, which allow for a more direct connection of cellular changes to a particular phenotype [5, 7]. Patient-specific hPSCs can be differentiated into retinal cell types such as photoreceptors and RPE in a consistent and reproducible manner to study retinal degenerative disorders that cause damage to more outer retinal cell types, with the remarkable ability to use such cells for drug screening, cell replacement, and targeted therapeutics [11, 12, 20–27]. Such studies have conducted thorough and in-depth experiments that have identified specific cellular changes in outer retinal cell types such as oxidative and ER stress, autophagy deficits, alterations in protein trafficking, and phagocytotic defects associated with disease-causing mutations. Disorders in which inner retinal cell types, such as RGCs, are the primary affected cell type remain less explored, with a limited number of studies available which describe cell-specific disease deficits and pathogenesis [14, 28–35]. As such, the use of hPSCs to study RGC-specific diseases is critical to unveiling disease mechanisms that cause the degeneration and eventual death of these cells in various optic neuropathies.

With the ability to elucidate mechanisms of neurodegeneration, which underlie numerous retinal diseases, hPSCs can be used to differentiate and enrich large populations of cells for drug development, including those high-throughput assays that utilize large chemical libraries to identify potential targets and pathways for drug development [4–6, 36]. This includes screening enriched populations of cells for safety and toxicity purposes, as well as developing new drugs which are able to target specific cellular pathways and circuits, with the hope of using such drugs for future therapeutics to treat neurodegenerative diseases [5, 6]. More specifically, patient-derived hPSCs have been differentiated into retinal cells such as RPE and photoreceptors and utilized to effectively screen drugs for future therapeutic purposes [20, 25, 27, 37] with studies using hPSC-derived RGCs for drug screening applications remaining largely unexplored [14, 38]. Although these strategies are not targeted at late-stage disease progression, the development of these approaches will be effective in early and mid stages of progression where few retinal neurons have degenerated and those remaining neurons can be rescued by pharmacological approaches.

Lastly, hPSCs can be utilized for the development of cellular replacement strategies, with these approaches often targeting late-stage disease progression where the majority of the cell population has degenerated, and as such, cellular replacement is no longer an option. Cell replacement strategies require the ability to derive specific cell types in a timely and reproducible manner as well as the capacity to identify a given cell type in living cultures. hPSC replacement strategies have been most successful for those cell types with relatively short-distance synaptic targets such as cells of the outer retina including photoreceptors and RPE [17–19, 21, 39]. These studies have generated functional retinal cell types with the ability for these cells to perform as they normally would within the innate retina [11, 19, 26], allowing for successful transplantation following late-stage diagnosis of outer retinal diseases. However, cell replacement for projection neurons of the retina such as RGCs presents numerous difficulties, as these cells will need to extend axons over long distances and navigate through an unhealthy environment to connect with their proper synaptic targets in the brain. In order for hPSC-derived RGCs to be used as an effective cell replacement strategy, many obstacles remain including proper integration into the retina, the extension of long neurites that can find their appropriate synaptic targets, and the formation of proper and functional synapses in the brain.

5.2 RGC Differentiation from hPSCs

In order to efficiently derive RGCs from hPSCs, a clear understanding of retinal development is essential, as many of the same principles are translated from development in vivo to inform in vitro systems. For instance, RGCs are one of the earliest born cell types in the retina, followed closely by amacrine cells, cones, and horizontal cells. At later stages of retinal development, rods are generated, followed by bipolar cells and eventually Muller glia. Similarly, when hPSCs are directed to differentiate toward a retinal lineage, a similar temporal sequence of development is recapitulated, with RGCs being one of the first cell types specified, followed sequentially by later-born retinal cell types as predicted from in vivo studies [8, 10, 14, 20, 40–42].

RGCs serve as the final output of the retina by sending light to brain regions such as the lateral geniculate nucleus or superior colliculus [43, 44]. As such, RGCs tend to have much larger cell bodies and thicker axons, both of which are needed for long-distance propagation of action potentials [45–47]. These axons fasciculate together in the nerve fiber layer and form the optic nerve, which relays information to multiple brain regions. hPSC-derived RGCs have been known to display similar distinct morphologies (Fig. 5.1), with long neurites, fasciculated axons, and large three-dimensional cell bodies [14, 31, 32, 48–50]. These RGCs have also shown some degree of target specificity, with neurites preferentially targeting superior colliculus explants in vitro [51]. Additionally, the RGC layer occupies a distinct position in the retinal architecture, residing in the innermost layers [52–54]. Similarly, as hPSCs have shown the ability to self-organize into optic cup-like structures called retinal organoids that recapitulate in vivo retinal organization, RGCs reside within the innermost layer of these structures and photoreceptors occupy the more peripheral layers [8, 10, 11, 14].

Within the retina, glutamate is used as the main excitatory neurotransmitter, and similar to their in vivo counterparts, it has been shown that hPSC-derived

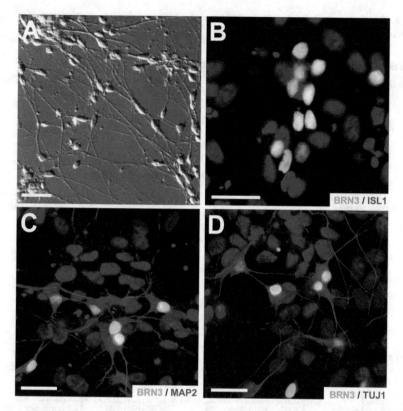

Fig. 5.1 Common markers of hPSC-derived retinal ganglion cell. hPSC-derived RGCs exhibit transcriptional and morphological features. (**a**) DIC imaging of retinal cultures demonstrated RGC-like morphology with large, three-dimensional somas, and long neurite projections. (**b, c**) hPSC-derived RGCs can be identified by the expression of RGC-associated transcription factors such as ISL1 and BRN3. (**c, d**) Immunocytochemistry displayed unique morphological features of RGCs with BRN3-positive cells extending lengthy MAP2- and TUJ1-positive neurites. Scale bar for (**a**) is 50 and 100 μm for (**b–d**)

RGCs have the ability to respond to glutamate, recapitulating the presynaptic organization in vivo [55, 56]. These cells have also exhibited EPSCs [31], action potentials [14, 31, 56, 57], spontaneous calcium transients [32], and are sensitive to the voltage-gated potassium and sodium channel blockers TEA and TTX, respectively [14, 57]. While these features do not definitively distinguish RGCs from other neuronal cell types, RGCs are the predominant cell type within the retina that have the capacity to both respond to glutamatergic stimulation and conduct action potentials [58].

In order to identify RGCs apart from other cell types in differentiating cultures of hPSCs, a variety of genetic markers are often used to confirm the identity of these cells (Table 5.1). Early studies utilized TUJ1 as a common marker of RGC-like cells [40, 55], which is expressed by RGCs but does not necessarily confer any specificity, as many projection neurons throughout the central nervous system also

Table 5.1 A summary of studies to date outlining the differentiation of retinal ganglion cells from human pluripotent stem cells

Authors	ihc markers	Functional characteristics	Disease state modeled	References
Huang	Brn3 Math 5 SNCG Islet-1 Thy1 TUJ1	Spiking activity, EPSCs	N/A	Riazifar [31], Chen [28]
Fingert	Math5 NF200 Thy1	N/A	TBK1-Normal Tension Glaucoma	Tucker [33]
Zack	Brn3 Islet1 TUJ1 Map2 NEFH NeuN Pax6 RBPMS SNCG Tau Thy1	Action potentials, responsive to AMPA/ Kainate	N/A	Sluch et al. [56], Sluch [59]
Takahashi	Brn3 SMI312 TUJ1 Thy1	N/A	N/A	Maekawa et al. [49], Kobayashi et al. [30]
Meyer	Brn3 Islet1 Map2 Tau HuC/D RBPMS Melanopsin Pax6 CART CDH6 FSTL4 SPP1 CB2 DCX	Action potentials, voltage-gated channels	E50K-normal tension glaucoma	Ohlemacher [14], Langer [48]
Wong	Brn3 Thy1 NEFM HuC/D TUJ1	Voltage-gated channels, Axonal transport	Leber's Hereditary Optic Neuropathy	Gill et al. [57], Wong [34]
Ahmad	Brn3 TUJ1 Thy1	Voltage-gated channels, action potentials, calcium transients	Six6-primary open-angle glaucoma	Teotia et al. [32, 51]

(continued)

Table 5.1 (continued)

Authors	ihc markers	Functional characteristics	Disease state modeled	References
Azuma	Brn3 Math5 Pax6 Islet1 TUJ1 SNCG Tau NFL/M NFH	Voltage-gated channels, action potentials	N/A	Tanaka et al. [50], Yokoi et al. [35]
Ge	Brn3 Islet1 Map2 Math5 NF200 Thy1	N/A	N/A	Huang et al. [29]

express this marker [60, 61]. Therefore, more specific markers were needed in the field to identify RGCs when derived from hPSCs.

The most common transcription factor used to identity RGCs has been BRN3 (Fig. 5.1). BRN3 is expressed specifically by RGCs within the retina and is expressed shortly after RGC specification persisting into adulthood [62–64]. However, BRN3 expression is also present in other cells including somatosensory and auditory neurons [62, 63, 65, 66]. Thus, when starting with a population of hPSCs that can differentiate into any cell type of the body, caution must be taken not to rely solely on one marker as proof of identity. Instead, the combinatorial expression of genetic markers, morphological features, and functional characteristics must be combined to definitively identify a presumptive RGC in vitro.

A variety of additional markers have been used to identify RGCs, including ISLET1, HuC/D, and SNCG. Within the retina, these markers show a high degree of specificity for RGCs but are less reliable in identifying hPSC-derived RGCs as the expression of these markers can be found in other neuronal cell types [67–70]. More recently, RNA-binding protein with multiple splicing (RBMPS) has been shown to specifically label RGCs, although this expression often occurs later in development, with limited use of this protein to identify RGCs early in hPSC differentiation [71].

Current research regarding RGC development and maturation has highlighted the diverse nature of these cells, with the discovery and identification of more than 30 subtypes that differ in molecular, morphological, and physiological properties in animal models [72, 73]. These subtypes express specific molecular markers which categorize RGCs into distinct subtypes. The identification of RGC subtypes further enhances the understanding of RGC characteristics including their molecular signatures as well as their mosaicism in the retina, with recent studies identifying a number of RGC subtypes in hPSC-derived cells [48]. This ability to identify hPSC-derived RGC subtypes will allow for the ability to tailor future studies identifying RGC subtypes in human-derived cells.

5.3 Translational Applications of hPSC-Derived RGCs

Human pluripotent stem cells provide a fundamental and unique tool in the study of human retinal degenerative diseases, including optic neuropathies such as glaucoma, which explicitly target RGCs. As these cells provide the critical connection between the eye and the brain to transmit visual information, their degeneration results in vision loss and eventual blindness. Traditionally, the study of disease states has been limited to the development and use of animal models, which have led to significant advances in understanding retinal disease progression but sometimes fail when translated to humans in a clinical setting due to important differences between species [74, 75].

In order to address these obstacles, researchers have focused their work on the use of human stem cells to study the pathogenesis and treatment of human degenerative diseases. The development of hPSCs revolutionized the field of disease modeling with the ability to generate cells from patient-specific sources with disease-causing mutations [2, 3, 76]. Furthermore, the use of hPSCs allows for the high-throughput screening of thousands of compounds for their therapeutic efficacy, as well as providing critical safety and toxicity information in human cells to drug developers that cannot be fully elucidated with animal models.

Many hPSC-based genetic models of RGC degeneration, specifically glaucomatous degeneration, have been developed over the past few years [14, 32, 33], including mutations in SIX6, TANK Binding Kinase 1 (TBK1), and Optineurin (OPTN), all of which have been associated with forms of familial normal tension glaucoma (NTG). SIX6 is widely known for its role in eye development and morphogenesis, although mutations in this transcription factor have also been implicated as a contributor to NTG [77, 78]. Interestingly, hPSCs derived from patients with a missense mutation in SIX6 generated neural and retinal cells inefficiently, with reduced expression and dysregulation of key developmental genes [32]. In addition, SIX6-mutant patient RGCs displayed severe developmental, morphological, and electrophysiological deficits, with reduced neurite outgrowth and deficiency in the expression of axonal guidance molecules. Furthermore, SIX6-mutated RGCs demonstrated significantly higher levels of activated caspase-3. It is hypothesized that this missense mutation in SIX6 results in developmentally defective RGCs that might put these RGCs at higher risk for degeneration in adulthood.

When hPSCs are derived from a patient population with a known genetic basis of underlying retinal disease, the resulting cells recreate certain features of the disease phenotype and model the degeneration associated with retinal diseases. Duplications in TBK1 have been associated with development of NTG, although its exact role remains poorly understood. It is hypothesized that due to the close association of TBK1 with autophagy, duplications could disrupt the autophagic pathway, leading to RGC degeneration [79–82]. Interestingly, the generation of hPSCs from patient sources with the TBK1 duplication demonstrated decreased levels of autophagy activation when compared to control RGCs, thereby allowing for subsequent studies of disease mechanisms leading to degeneration of RGCs [33]. The accumulated results of studies such as these provide an important stepping stone towards the development of therapeutic approaches for retinal degenerative diseases.

The directed differentiation of RGCs from hPSCs provides a large quantity of cells for drug screening efforts, which can then specifically target RGCs to assess the ability of candidate compounds to rescue a disease phenotype. hPSC-derived RGCs allow for the screening of new drug compounds as well as the development of personalized treatment therapies, particularly when derived from patient-specific sources with a confirmed degenerative phenotype. With this in mind, recent studies have successfully recapitulated some of the degenerative process associated with optic neuropathies in hPSC-derived RGCs, with subsequent drug screening approaches identifying candidate treatment factors capable of rescuing RGC degeneration. Mutations in the OPTN protein have been associated with multiple types of neurodegeneration, including glaucoma and amyotrophic lateral sclerosis and glaucoma [38, 83]. The E50K missense mutation in OPTN has been associated with severe and early-onset NTG in a clinical setting [84]. Like TBK1, OPTN is closely associated with the autophagy pathway as it can act as an autophagy receptor [85, 86]. It is thought that OPTN mutations lead to degeneration through the dysregulation or blockage of the autophagy pathway. In a recent study, hPSC-derived RGCs derived from a patient with an E50K mutation in the OPTN gene displayed elevated levels of activated caspase-3 compared to control lines, with the ability to rescue these damaged RGCs following the treatment of these cells with neuroprotective factors such as BDNF and GDNF [14]. Therefore, the use of patient-derived hPSCs has provided an in-depth understanding of disease progression and mechanisms, which have subsequently enabled the identification of compounds to combat the degeneration of RGCs.

Postautosomal dominant optic atrophy (DOA) is the most common hereditary optic atrophy that culminates in degeneration of RGCs and eventual central vision loss [87]. Mutations in the OPA1 gene, which affect inner mitochondrial membrane proteins, are the most common cause of DOA, resulting in mitochondrial dysfunction, decreased ATP production, as well as mitochondrial fragmentation [88–90]. Consequently, generation of hPSCs from patients carrying an OPA1 mutation exhibited significantly more apoptosis and inefficiently differentiated into RGCs, suggesting that mutations in OPA1 mediate apoptosis and contribute to the pathogenesis of optic atrophy. In addition to modeling a disease, hPSC-derived RGCs can also be used to uncover neuroprotective agents that slow or even halt the progression of RGC degeneration. The addition of Noggin and estrogen to OPA1-mutated iPSCs promoted the differentiation of RGCs, representing potential therapeutic agents for OPA1-related optic atrophy. Taken together, these studies represent the first demonstration of disease modeling using hPSC-derived RGCs and are an important step forward in understanding disease mechanisms and identifying potential therapeutic interventions.

Traditional approaches to combat RGC degeneration have relied on treatment during early stages of the disease process when neuroprotection is still feasible. hPSC-derived RGCs can be used to effectively study early stages of retinal degenerative diseases, and subsequent drug screening approaches may aid in the development of neuroprotective strategies for degenerating RGCs. Unfortunately, a majority of optic neuropathy patients have experienced significant RGC loss by the time of diagnosis, and the large number of cells that are irreversibly lost at later stages in the disease renders neuroprotection futile. In such cases, the transplantation of healthy RGCs to replace degenerated cells remains the final option to restore visual function. However, there has been a lack of successful development of replacement strategies for RGCs due to obstacles such as the long-distance projection of RGC axons as well as the functional formation of synapses with appropriate postsynaptic targets. In spite of these obstacles, studies have demonstrated the ability of hPSC-derived RGCs to survive on a tissue-engineered scaffold following transplantation into the vitreous chamber of rabbits and rhesus monkeys [91].

Recent work has described the ability to elegantly combine the use of CRISPR gene editing and hPSC technology to elucidate potential therapeutic pathways involved in RGC degeneration and find suitable compounds to intervene within this pathway [36, 59]. In such studies, hPSC-derived RGCs have engineered to express a tdTomato-P2A-Thy1.2 reporter driven under the expression of RGC marker BRN3, with the use of the Thy1.2 surface receptor to further immunopurify and isolate RGCs. The resultant tdTomato-positive RGCs were treated with colchicine to simulate axonal injury allowing for the subsequent investigation of pathways that mediated RGC death. In particular, the dual leucine zipper kinase (DLK) pathway and its downstream partner leucine zipper kinase (LZK) pathway were discovered as mediators of RGC cell death and were proposed as possible targets for intervention to increase cell survival. Treatment of RGCs with Sunitinib, a FDA-approved drug that is known to interfere with the DLK and LZK pathways, was shown to enhance survival of injured hPSC-RGCs in a dose-dependent manner. This study provided integral insights into RGC degeneration, particularly the role of the DLK and LZK pathway in RGC injury, and demonstrated the first use of CRISPR-engineered hPSC-derived RGCs for drug screening applications.

Taken together, the generation of genetic disease models from hPSC-derived RGCs allows for the previously unattainable insight into early disease mechanisms and the development of tests for early detection in a clinical setting. In addition, hPSCs also allow for the generation of large populations of patient-specific cells for high-throughput drug screening. Finally, the derivation of patient-specific cells is a groundbreaking advancement in personalized medicine as the ability to screen compounds on a patient's own cells could greatly optimize the process of discovering the most effective therapies for retinal degeneration.

5.4   Future Applications of hPSC-Derived RGCs

Efforts from the past decade have utilized hPSCs to study the development and disease pathologies of all different cell types of the retina. While most efforts have emphasized cells of the outer retina, the use of hPSCs to study RGC-specific development and disease is largely lacking, with important implications for future studies elucidating developmental and disease mechanisms within such cells. Future applications of hPSC-derived RGCs include the use of retinal organoids as a model for RGC development and disease, exploring RGC subtypes in human cells and how these subtypes may be differentially affected in disease states, studying cell replacement and drug screening therapies for treating optic neuropathies, and utilizing genome editing to enhance the study of human RGCs in vitro.

The term retinal organoid refers to a three-dimensional structure derived from stem cells, which recapitulates the temporal development and spatial lamination of the retina. As RGCs are one of the first cells to develop, retinal organoids display similar lamination to the in vivo retina, with RGCs found within inner layers and photoreceptors residing in more peripheral layers (Fig. 5.2). Recent efforts have utilized retinal organoids for studying cells and diseases of the outer retina including photoreceptors and RPE due to their short synaptic contacts and the ability for these cells to mature in this environment [11, 16, 26, 92]. However, there have been a limited number of studies focusing on RGC development within retinal organoids. As RGCs are one of the first to develop within the retina, the accessibility of studying RGC differentiation within retinal organoids provides a more feasible timeline than that of photoreceptors that take upward of 200 days to become fully mature. Although the timeline of development is considerably shorter, the ability for RGCs to fully mature into a bonafide cell type remains unclear. As the projection neurons of the retina, RGCs extend long axons out of the eye and into the brain to synapse

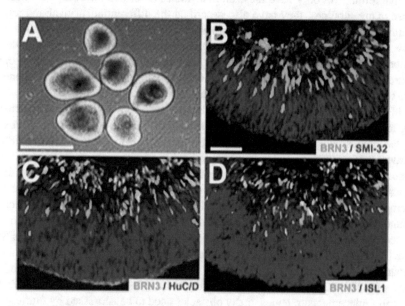

Fig. 5.2 Retinal organoids sustain unique morphology and cellular lamination. (**a**) Retinal organoids exhibit a bright outer ring around the periphery indicating retinal organization and lamination. (**b–d**) Retinal organoids exhibit widespread expression of the RGC marker BRN3, neural markers HUC/D, and ISLET1, as well as cytoskeletal marker SMI-32 co-localized within apical layers of the organoid. Scale bars equal 500 μm for (**a**) and 50 μm for (**b–d**)

with postsynaptic targets. Within retinal organoids, RGC axons are limited to the space and confinement of these structures, without the current accessibility to extend outward toward a specific target. Additionally, RGCs are one of the only cell types in the retina to fire action potentials in order to conduct their visual information to the brain. No studies currently exist displaying RGCs within retinal organoids to possess this capability, which is crucial for using cells grown in such a manner for reliable developmental and disease studies, although dissociated hPSC-derived RGCs have demonstrated functional properties [14, 31, 32, 51, 56, 59]. Future studies in this area may find the need to incorporate biomechanical engineering approaches in order to devise an exit for RGC axons confined within retinal organoids using an extracellular matrix mold or scaffolding device. Additionally, future studies will need to focus on expediting the maturation of RGCs within retinal organoids, including their ability to fire action potentials, in order for these cells to be used as a model which closely recapitulates the mechanisms of the human retina.

In recent years, the study of RGCs has become more complex with the discovery of more than 30 different subtypes of these cells, all of which possess varying molecular, structural, and functional characteristics [72, 73]. Although the majority of RGC subtypes have been characterized in animal models, more recently a variety of these subtypes have also been identified in hPSC-derived cells [48]. Future efforts will focus on the differentiation of specific subtypes from hPSCs as well as their ability to conduct similar functional characteristics in vitro as observed within the in vivo retina. Not only have these animal studies described the complex nature of these many subtypes, they have also described the differential survival and regeneration of different subtypes following acute injury or disease [93–99]. As such, this leaves opportunity to study this phenomenon within hPSC-derived RGCs in order to properly understand the degeneration of specific RGC subtypes in different optic neuropathies as well as provide a greater understanding of how to address degenerative cell loss in cell replacement studies and therapeutics in the future.

hPSC-derived RGCs also have the ability to be used for large-scale drug screening to address early disease progression as well as provide an opportunity for cell replacement strategies to address late-stage degeneration [5, 6]. In the future, hPSC-derived RGCs can be grown in large, reproducible quantities from numerous patient sources harboring disease mutations to utilize large chemical libraries and databases, with the hope of finding specific compounds that may provide rescue and therapeutic benefits to diseased RGCs. More so, the discovery of therapeutic compounds could be utilized for early disease diagnosis and progression in order to halt or rescue cell-specific deficits. To address late-stage disease progression, hPSCs provide an advantageous source for cell replacement strategies as these cells can be expanded indefinitely and differentiated into all cell types of the body, including RGCs [14, 31, 32, 35, 57, 59, 100]. In order to be able to use hPSC-derived RGCs for cell replacement strategies, many obstacles need to be addressed by future studies. To be used for cell replacement, hPSC-derived RGCs must possess the capacity to differentiate and function as an in vivo RGC would, including the ability fire repetitive action potentials, extend axons long distances to a proper target, and make functional synaptic connections. Each of these points will need to be addressed fully by future studies in order for hPSCs to be used as a reliable source for cell replacement in end-stage disease progression affecting RGCs.

Lastly, CRIPSR engineering is an emerging and fast-growing technology which can be used to better study RGC development and disease at high and in-depth capacities in future studies [101]. To study early RGC development, CRISPR technology has the ability to engineer transcriptional activators or repressors [102–104]. These allow for the conditional expression of genes at specific time points in development, with the ability to identify pathways which are essential for proper RGC maturation and may be dysfunctional in disease states. hPSC-derived RGCs can also be engineered to express epitope tags that allow for visualization of proteins that are especially difficult to study or for which no antibodies exist, including those involving apoptotic and autophagy pathways [105, 106]. More so, this technique would also allow a fluorescent reporter to tag a protein of interest, allowing for the identification and study of proteins in living cultures, including the ability to easily identify RGCs in vitro and study their electrophysiological properties [56, 59, 102, 105].

CRISPR engineering can also be used to create specific gene mutations [101, 107], including those known to cause various optic neuropathies including mutations in TBK1, OPTN, and SIX6. This technology allows for the generation of disease-harboring hPSCs for rare genetic determinants where patient sources are scarce and allows for the direct creation of an isogenic control which is important for identifying disease phenotypes across many cell lines. More so, CRISPR technology can be used to correct gene mutations in cases where patient samples are abundant and hPSCs can be readily reprogrammed and differentiated. The ability to correct these specific gene mutations allows for a more direct connection of disease mechanisms and phenotypes between healthy and diseased samples. CRISPR-engineered hPSCs also provide a personalized source of cells for cell replacement therapy with the ability to collect somatic cells from patients, correct gene-causing mutations, reprogram them into healthy hPSCs, and use patient-specific cells for replacement of degenerated RGCs [108, 109]. CRISPR engineering will be essential and necessary for future studies of hPSC-derived RGCs in regard to studying their proper development in vitro, how they are affected in disease states, and how these cells can be used for cell replacement strategies.

5.5   Conclusions

Over the past decade, hPSC technology has been utilized as a reliable tool to elucidate cell development and various neurodegenerative diseases when derived from patient-specific sources. hPSCs can be readily differentiated into all cell types of the retina in a manner which closely recapitulates retinogenesis, with a variety of cell types exhibiting characteristics of bonafide retinal cell types such as photoreceptors and RPE. The majority of these studies have observed outer retinal cell development and disease, with limited studies available looking at RGC differentiation and maturation and diseases that target RGCs such as glaucoma. A greater understanding of RGC-specific differentiation and maturation is needed in order to elucidate important development pathways and signaling cascades, which may be adversely affected in disease which target RGCs. CRISPR engineering has opened up the possibilities of studying the development of RGCs in vitro by the insertion of fluorescent reporters driven by important neural, retinal, and RGC-specific genes, allowing a more comprehensive technique for studying cell-specific differentiation. An enhanced understanding of RGC development will create the opportunity to develop a cell type that closely resembles native RGCs and can be used as a more reliable model for studying disease phenotypes as well as their ability to be used for cell replacement therapies for optic neuropathies in the future.

Acknowledgments Grant support was provided by the National Eye Institute (R01 EY024984 to JSM), Indiana Department of Health Brain and Spinal Cord Injury Fund (JSM), an IU Collaborative Research Grant from the Office of the Vice President for Research (JSM), an award from the IU Signature Center for Brain and Spinal Cord Injury (JSM), a grant from Stark Neurosciences Research Institute, Eli Lilly and Company, and by the Indiana Clinical and Translational Sciences Institute, funded in part by grant # UL1TR001108 from the National Institutes of Health, National Center for Advancing Translational Sciences (SO) and an IUPUI Graduate Office First Year University Fellowship (KL), and the Purdue Research Foundation Fellowship (KL).

Author Contributions: SO, KL, CF, ME, JM: manuscript writing; SO, KL, CF, EF: Data collection and figure design; SO, KL, CF, JM: manuscript revisions, JM: final approval of manuscript.

References

1. Harwerth RS, Quigley HA (2006) Visual field defects and retinal ganglion cell losses in patients with glaucoma. Arch Ophthalmol 124:853–859
2. Takahashi K, Tanabe K, Ohnuki M, Narita M, Ichisaka T, Tomoda K, Yamanaka S (2007) Induction of pluripotent stem cells from adult human fibroblasts by defined factors. Cell 131:861–872
3. Yu J, Vodyanik MA, Smuga-Otto K, Antosiewicz-Bourget J, Frane JL, Tian S, Nie J, Jonsdottir GA, Ruotti V, Stewart R et al (2007) Induced pluripotent stem cell lines derived from human somatic cells. Science 318:1917–1920
4. Ebert AD, Svendsen CN (2010) Human stem cells and drug screening: opportunities and challenges. Nat Rev Drug Discov 9:367–372
5. Grskovic M, Javaherian A, Strulovici B, Daley GQ (2011) Induced pluripotent stem cells--opportunities for disease modelling and drug discovery. Nat Rev Drug Discov 10:915–929
6. Inoue H, Yamanaka S (2011) The use of induced pluripotent stem cells in drug development. Clin Pharmacol Ther 89:655–661
7. Marchetto MC, Brennand KJ, Boyer LF, Gage FH (2011) Induced pluripotent stem cells (iPSCs) and neurological disease modeling: progress and promises. Hum Mol Genet 20:R109–R115
8. Eiraku M, Takata N, Ishibashi H, Kawada M, Sakakura E, Okuda S, Sekiguchi K, Adachi T, Sasai Y (2011) Self-organizing optic-cup morphogenesis in three-dimensional culture. Nature 472:51–56
9. Meyer JS, Shearer RL, Capowski EE, Wright LS, Wallace KA, McMillan EL, Zhang SC, Gamm DM (2009) Modeling early retinal development with human embryonic and induced pluripotent stem cells. Proc Natl Acad Sci U S A 106:16698–16703

10. Nakano T, Ando S, Takata N, Kawada M, Muguruma K, Sekiguchi K, Saito K, Yonemura S, Eiraku M, Sasai Y (2012) Self-formation of optic cups and storable stratified neural retina from human ESCs. Cell Stem Cell 10:771–785

11. Zhong X, Gutierrez C, Xue T, Hampton C, Vergara MN, Cao LH, Peters A, Park TS, Zambidis ET, Meyer JS et al (2014) Generation of three-dimensional retinal tissue with functional photoreceptors from human iPSCs. Nat Commun 5:4047

12. Kamao H, Mandai M, Okamoto S, Sakai N, Suga A, Sugita S, Kiryu J, Takahashi M (2014) Characterization of human induced pluripotent stem cell-derived retinal pigment epithelium cell sheets aiming for clinical application. Stem Cell Rep 2:205–218

13. Ohlemacher SK, Iglesias CL, Sridhar A, Gamm DM, Meyer JS (2015) Generation of highly enriched populations of optic vesicle-like retinal cells from human pluripotent stem cells. Curr Protoc Stem Cell Biol 32:1h.8.1–1h.820

14. Ohlemacher SK, Sridhar A, Xiao Y, Hochstetler AE, Sarfarazi M, Cummins TR, Meyer JS (2016) Stepwise Differentiation of Retinal Ganglion Cells from Human Pluripotent Stem Cells Enables Analysis of Glaucomatous Neurodegeneration. Stem Cells 34:1553–1562

15. Reichman S, Slembrouck A, Gagliardi G, Chaffiol A, Terray A, Nanteau C, Potey A, Belle M, Rabesandratana O, Duebel J et al (2017) Generation of storable retinal organoids and retinal pigmented epithelium from adherent human iPS cells in xeno-free and feeder-free conditions. Stem Cells 35:1176–1188

16. Volkner M, Zschatzsch M, Rostovskaya M, Overall RW, Busskamp V, Anastassiadis K, Karl MO (2016) Retinal organoids from pluripotent stem cells efficiently recapitulate retinogenesis. Stem Cell Rep 6:525–538

17. Carr AJ, Vugler AA, Hikita ST, Lawrence JM, Gias C, Chen LL, Buchholz DE, Ahmado A, Semo M, Smart MJ et al (2009) Protective effects of human iPS-derived retinal pigment epithelium cell transplantation in the retinal dystrophic rat. PLoS One 4:e8152

18. da Cruz L, Fynes K, Georgiadis O, Kerby J, Luo YH, Ahmado A, Vernon A, Daniels JT, Nommiste B, Hasan SM et al (2018) Phase 1 clinical study of an embryonic stem cell-derived retinal pigment epithelium patch in age-related macular degeneration. Nat Biotechnol 36:328

19. Lamba DA, McUsic A, Hirata RK, Wang PR, Russell D, Reh TA (2010) Generation, purification and transplantation of photoreceptors derived from human induced pluripotent stem cells. PLoS One 5:e8763

20. Meyer JS, Howden SE, Wallace KA, Verhoeven AD, Wright LS, Capowski EE, Pinilla I, Martin JM, Tian S, Stewart R et al (2011) Optic vesicle-like structures derived from human pluripotent stem cells facilitate a customized approach to retinal disease treatment. Stem Cells 29:1206–1218

21. Kruczek K, Gonzalez-Cordero A, Goh D, Naeem A, Jonikas M, Blackford SJI, Kloc M, Duran Y, Georgiadis A, Sampson RD et al (2017) Differentiation and transplantation of embryonic stem cell-derived cone photoreceptors into a mouse model of end-stage retinal degeneration. Stem Cell Rep 8:1659–1674

22. Maruotti J, Wahlin K, Gorrell D, Bhutto I, Lutty G, Zack DJ (2013) A simple and scalable process for the differentiation of retinal pigment epithelium from human pluripotent stem cells. Stem Cells Transl Med 2:341–354

23. Mellough CB, Sernagor E, Moreno-Gimeno I, Steel DH, Lako M (2012) Efficient stage-specific differentiation of human pluripotent stem cells toward retinal photoreceptor cells. Stem Cells 30:673–686

24. Peng CH, Huang KC, Lu HE, Syu SH, Yarmishyn AA, Lu JF, Buddhakosai W, Lin TC, Hsu CC, Hwang DK et al (2018) Generation of induced pluripotent stem cells from a patient with X-linked juvenile retinoschisis. Stem Cell Res 29:152–156

25. Singh R, Kuai D, Guziewicz KE, Meyer J, Wilson M, Lu J, Smith M, Clark E, Verhoeven A, Aguirre GD et al (2015) Pharmacological modulation of photoreceptor outer segment degradation in a human iPS cell model of inherited macular degeneration. Mol Ther 23:1700–1711

26. Wahlin KJ, Maruotti JA, Sripathi SR, Ball J, Angueyra JM, Kim C, Grebe R, Li W, Jones BW, Zack DJ (2017) Photoreceptor outer segment-like structures in long-term 3D retinas from human pluripotent stem cells. Sci Rep 7:766

27. Yoshida T, Ozawa Y, Suzuki K, Yuki K, Ohyama M, Akamatsu W, Matsuzaki Y, Shimmura S, Mitani K, Tsubota K et al (2014) The use of induced pluripotent stem cells to reveal pathogenic gene mutations and explore treatments for retinitis pigmentosa. Mol Brain 7:45

28. Chen J, Riazifar H, Guan MX, Huang T (2016) Modeling autosomal dominant optic atrophy using induced pluripotent stem cells and identifying potential therapeutic targets. Stem Cell Res Ther 7:2

29. Huang L, Chen M, Zhang W, Sun X, Liu B, Ge J (2018) Retinoid acid and taurine promote NeuroD1-induced differentiation of induced pluripotent stem cells into retinal ganglion cells. Mol Cell Biochem 438:67–76

30. Kobayashi W, Onishi A, Tu HY, Takihara Y, Matsumura M, Tsujimoto K, Inatani M, Nakazawa T, Takahashi M (2018) Culture systems of dissociated mouse and human pluripotent stem cell-derived retinal ganglion cells purified by two-step immunopanning. Invest Ophthalmol Vis Sci 59:776–787

31. Riazifar H, Jia Y, Chen J, Lynch G, Huang T (2014) Chemically induced specification of retinal ganglion cells from human embryonic and induced pluripotent stem cells. Stem Cells Transl Med 3:424–432

32. Teotia P, Van Hook MJ, Wichman CS, Allingham RR, Hauser MA, Ahmad I (2017b) Modeling glaucoma: retinal ganglion cells generated from induced pluripotent stem cells of patients with SIX6 risk allele show developmental abnormalities. Stem Cells 35:2239–2252

33. Tucker BA, Solivan-Timpe F, Roos BR, Anfinson KR, Robin AL, Wiley LA, Mullins RF, Fingert JH (2014) Duplication of TBK1 stimulates autophagy in iPSC-derived retinal cells from a patient with normal tension glaucoma. J Stem Cell Res Ther 3:161

34. Wong RCB, Lim SY, Hung SSC, Jackson S, Khan S, Van Bergen NJ, De Smit E, Liang HH, Kearns LS, Clarke L et al (2017) Mitochondrial replacement in an iPSC model of Leber's hereditary optic neuropathy. Aging 9:1341–1350

35. Yokoi T, Tanaka T, Matsuzaka E, Tamalu F, Watanabe SI, Nishina S, Azuma N (2017) Effects of neuroactive agents on axonal growth and pathfinding of retinal ganglion cells generated from human stem cells. Sci Rep 7:16757

36. Welsbie DS, Mitchell KL, Jaskula-Ranga V, Sluch VM, Yang Z, Kim J, Buehler E, Patel A, Martin SE, Zhang PW et al (2017) Enhanced functional genomic screening identifies novel mediators of dual leucine zipper kinase-dependent injury signaling in neurons. Neuron 94:1142–1154.e1146

37. Jin ZB, Okamoto S, Osakada F, Homma K, Assawachananont J, Hirami Y, Iwata T, Takahashi M (2011) Modeling retinal degeneration using patient-specific induced pluripotent stem cells. PLoS One 6:e17084

38. Minegishi Y, Nakayama M, Iejima D, Kawase K, Iwata T (2016) Significance of optineurin mutations in glaucoma and other diseases. Prog Retin Eye Res 55:149–181

39. Assawachananont J, Mandai M, Okamoto S, Yamada C, Eiraku M, Yonemura S, Sasai Y, Takahashi M (2014) Transplantation of embryonic and induced pluripotent stem cell-derived 3D retinal sheets into retinal degenerative mice. Stem Cell Rep 2:662–674

40. Hirami Y, Osakada F, Takahashi K, Okita K, Yamanaka S, Ikeda H, Yoshimura N, Takahashi M (2009) Generation of retinal cells from mouse and human induced pluripotent stem cells. Neurosci Lett 458:126–131

41. Sridhar A, Ohlemacher SK, Langer KB, Meyer JS (2016) Robust differentiation of mRNA-reprogrammed human induced pluripotent stem cells toward a retinal lineage. Stem Cells Transl Med 5:417–426

42. Sridhar A, Steward MM, Meyer JS (2013) Nonxenogeneic growth and retinal differentiation of human induced pluripotent stem cells. Stem Cells Transl Med 2:255–264

43. Herrera E, Erskine L, Morenilla-Palao C (2017) Guidance of retinal axons in mammals. Semin Cell Dev Biol 85:48

44. Martersteck EM, Hirokawa KE, Evarts M, Bernard A, Duan X, Li Y, Ng L, Oh SW, Ouellette B, Royall JJ et al (2017) Diverse central projection patterns of retinal ganglion cells. Cell Rep 18:2058–2072

45. Kolb H, Nelson R, Mariani A (1981) Amacrine cells, bipolar cells and ganglion cells of the cat retina: a Golgi study. Vision Res 21:1081–1114
46. Volgyi B, Chheda S, Bloomfield SA (2009) Tracer coupling patterns of the ganglion cell subtypes in the mouse retina. J Comp Neurol 512:664–687
47. Watanabe M, Rodieck RW (1989) Parasol and midget ganglion cells of the primate retina. J Comp Neurol 289:434–454
48. Langer KB, Ohlemacher SK, Phillips MJ, Fligor CM, Jiang P, Gamm DM, Meyer JS (2018) Retinal ganglion cell diversity and subtype specification from human pluripotent stem cells. Stem Cell Rep 10:1282–1293
49. Maekawa Y, Onishi A, Matsushita K, Koide N, Mandai M, Suzuma K, Kitaoka T, Kuwahara A, Ozone C, Nakano T et al (2016) Optimized culture system to induce neurite outgrowth from retinal ganglion cells in three-dimensional retinal aggregates differentiated from mouse and human embryonic stem cells. Curr Eye Res 41:558–568
50. Tanaka T, Yokoi T, Tamalu F, Watanabe S, Nishina S, Azuma N (2015) Generation of retinal ganglion cells with functional axons from human induced pluripotent stem cells. Sci Rep 5:8344
51. Teotia P, Van Hook MJ, Ahmad I (2017a) A co-culture model for determining the target specificity of the de novo generated retinal ganglion cells. Bio Protoc 7:e2212
52. Curcio CA, Allen KA (1990) Topography of ganglion cells in human retina. J Comp Neurol 300:5–25
53. Dowling JE (1970) Organization of vertebrate retinas. Invest Ophthalmol 9:655–680
54. Dowling JE, Boycott BB (1966) Organization of the primate retina: electron microscopy. Proc R Soc Lond B Biol Sci 166:80–111
55. Lamba DA, Karl MO, Ware CB, Reh TA (2006) Efficient generation of retinal progenitor cells from human embryonic stem cells. Proc Natl Acad Sci U S A 103:12769–12774
56. Sluch VM, Davis CH, Ranganathan V, Kerr JM, Krick K, Martin R, Berlinicke CA, Marsh-Armstrong N, Diamond JS, Mao HQ et al (2015) Differentiation of human ESCs to retinal ganglion cells using a CRISPR engineered reporter cell line. Sci Rep 5:16595
57. Gill KP, Hung SS, Sharov A, Lo CY, Needham K, Lidgerwood GE, Jackson S, Crombie DE, Nayagam BA, Cook AL et al (2016) Enriched retinal ganglion cells derived from human embryonic stem cells. Sci Rep 6:30552
58. Velte TJ, Masland RH (1999) Action potentials in the dendrites of retinal ganglion cells. J Neurophysiol 81:1412–1417
59. Sluch VM, Chamling X, Liu MM, Berlinicke CA, Cheng J, Mitchell KL, Welsbie DS, Zack DJ (2017) Enhanced stem cell differentiation and immunopurification of genome engineered human retinal ganglion cells. Stem Cells Transl Med 6:1972–1986
60. Sharma RK, Netland PA (2007) Early born lineage of retinal neurons express class III beta-tubulin isotype. Brain Res 1176:11–17
61. Sullivan KF, Cleveland DW (1986) Identification of conserved isotype-defining variable region sequences for four vertebrate beta tubulin polypeptide classes. Proc Natl Acad Sci U S A 83:4327–4331
62. Erkman L, McEvilly RJ, Luo L, Ryan AK, Hooshmand F, O'Connell SM, Keithley EM, Rapaport DH, Ryan AF, Rosenfeld MG (1996) Role of transcription factors Brn-3.1 and Brn-3.2 in auditory and visual system development. Nature 381:603–606
63. Gan L, Xiang M, Zhou L, Wagner DS, Klein WH, Nathans J (1996) POU domain factor Brn-3b is required for the development of a large set of retinal ganglion cells. Proc Natl Acad Sci U S A 93:3920–3925
64. Nadal-Nicolas FM, Jimenez-Lopez M, Sobrado-Calvo P, Nieto-Lopez L, Canovas-Martinez I, Salinas-Navarro M, Vidal-Sanz M, Agudo M (2009) Brn3a as a marker of retinal ganglion cells: qualitative and quantitative time course studies in naive and optic nerve-injured retinas. Invest Ophthalmol Vis Sci 50:3860–3868
65. Chambers SM, Qi Y, Mica Y, Lee G, Zhang XJ, Niu L, Bilsland J, Cao L, Stevens E, Whiting P et al (2012) Combined small-molecule inhibition accelerates developmental timing and converts human pluripotent stem cells into nociceptors. Nat Biotechnol 30:715–720

66. Koehler KR, Mikosz AM, Molosh AI, Patel D, Hashino E (2013) Generation of inner ear sensory epithelia from pluripotent stem cells in 3D culture. Nature 500:217–221
67. Ericson J, Thor S, Edlund T, Jessell TM, Yamada T (1992) Early stages of motor neuron differentiation revealed by expression of homeobox gene Islet-1. Science 256:1555–1560
68. Genead R, Danielsson C, Andersson AB, Corbascio M, Franco-Cereceda A, Sylven C, Grinnemo KH (2010) Islet-1 cells are cardiac progenitors present during the entire lifespan: from the embryonic stage to adulthood. Stem Cells Dev 19:1601–1615
69. Nguyen HN, Byers B, Cord B, Shcheglovitov A, Byrne J, Gujar P, Kee K, Schule B, Dolmetsch RE, Langston W et al (2011) LRRK2 mutant iPSC-derived DA neurons demonstrate increased susceptibility to oxidative stress. Cell Stem Cell 8:267–280
70. Phillips RJ, Hargrave SL, Rhodes BS, Zopf DA, Powley TL (2004) Quantification of neurons in the myenteric plexus: an evaluation of putative pan-neuronal markers. J Neurosci Methods 133:99–107
71. Rodriguez AR, de Sevilla Muller LP, Brecha NC (2014) The RNA binding protein RBPMS is a selective marker of ganglion cells in the mammalian retina. J Comp Neurol 522:1411–1443
72. Dhande OS, Stafford BK, Lim JA, Huberman AD (2015) Contributions of retinal ganglion cells to subcortical visual processing and behaviors. Ann Rev Vis Sci 1:291–328
73. Sanes JR, Masland RH (2015) The types of retinal ganglion cells: current status and implications for neuronal classification. Annu Rev Neurosci 38:221–246
74. Shanks N, Greek R, Greek J (2009) Are animal models predictive for humans? Philos Ethics Humanit Med 4:2
75. Shelley C (2012) Complex systems, evolution, and animal models. Stud Hist Philos Biol Biomed Sci 43:311
76. Park IH, Arora N, Huo H, Maherali N, Ahfeldt T, Shimamura A, Lensch MW, Cowan C, Hochedlinger K, Daley GQ (2008) Disease-specific induced pluripotent stem cells. Cell 134:877–886
77. Carnes MU, Liu YP, Allingham RR, Whigham BT, Havens S, Garrett ME, Qiao C, Katsanis N, Wiggs JL, Pasquale LR et al (2014) Discovery and functional annotation of SIX6 variants in primary open-angle glaucoma. PLoS Genet 10:e1004372
78. Wiggs JL, Yaspan BL, Hauser MA, Kang JH, Allingham RR, Olson LM, Abdrabou W, Fan BJ, Wang DY, Brodeur W et al (2012) Common variants at 9p21 and 8q22 are associated with increased susceptibility to optic nerve degeneration in glaucoma. PLoS Genet 8:e1002654
79. Fingert JH (2011) Primary open-angle glaucoma genes. Eye (Lond) 25:587–595
80. Kawase K, Allingham RR, Meguro A, Mizuki N, Roos B, Solivan-Timpe FM, Robin AL, Ritch R, Fingert JH (2012) Confirmation of TBK1 duplication in normal tension glaucoma. Exp Eye Res 96:178–180
81. Korac J, Schaeffer V, Kovacevic I, Clement AM, Jungblut B, Behl C, Terzic J, Dikic I (2013) Ubiquitin-independent function of optineurin in autophagic clearance of protein aggregates. J Cell Sci 126:580–592
82. Ryan TA, Tumbarello DA (2018) Optineurin: a coordinator of membrane-associated cargo trafficking and autophagy. Front Immunol 9:1024
83. Maruyama H, Morino H, Ito H, Izumi Y, Kato H, Watanabe Y, Kinoshita Y, Kamada M, Nodera H, Suzuki H et al (2010) Mutations of optineurin in amyotrophic lateral sclerosis. Nature 465:223–226
84. Aung T, Rezaie T, Okada K, Viswanathan AC, Child AH, Brice G, Bhattacharya SS, Lehmann OJ, Sarfarazi M, Hitchings RA (2005) Clinical features and course of patients with glaucoma with the E50K mutation in the optineurin gene. Invest Ophthalmol Vis Sci 46:2816–2822
85. Nixon RA (2013) The role of autophagy in neurodegenerative disease. Nat Med 19:983–997
86. Ying H, Yue BY (2016) Optineurin: the autophagy connection. Exp Eye Res 144:73–80
87. Chen T, Wu H, Guo L, Liu L (2015) A modified rife algorithm for off-grid DOA estimation based on sparse representations. Sensors 15:29721–29733

88. Barboni P, Savini G, Cascavilla ML, Caporali L, Milesi J, Borrelli E, La Morgia C, Valentino ML, Triolo G, Lembo A et al (2014) Early macular retinal ganglion cell loss in dominant optic atrophy: genotype-phenotype correlation. Am J Ophthalmol 158:628–636.e623

89. Lodi R, Tonon C, Valentino ML, Iotti S, Clementi V, Malucelli E, Barboni P, Longanesi L, Schimpf S, Wissinger B et al (2004) Deficit of in vivo mitochondrial ATP production in OPA1-related dominant optic atrophy. Ann Neurol 56:719–723

90. Zanna C, Ghelli A, Porcelli AM, Karbowski M, Youle RJ, Schimpf S, Wissinger B, Pinti M, Cossarizza A, Vidoni S et al (2008) OPA1 mutations associated with dominant optic atrophy impair oxidative phosphorylation and mitochondrial fusion. Brain 131:352–367

91. Li K, Zhong X, Yang S, Luo Z, Li K, Liu Y, Cai S, Gu H, Lu S, Zhang H et al (2017) HiPSC-derived retinal ganglion cells grow dendritic arbors and functional axons on a tissue-engineered scaffold. Acta Biomater 54:117–127

92. Phillips MJ, Wallace KA, Dickerson SJ, Miller MJ, Verhoeven AD, Martin JM, Wright LS, Shen W, Capowski EE, Percin EF et al (2012) Blood-derived human iPS cells generate optic vesicle-like structures with the capacity to form retinal laminae and develop synapses. Invest Ophthalmol Vis Sci 53:2007–2019

93. Daniel S, Clark AF, McDowell CM (2018) Subtype-specific response of retinal ganglion cells to optic nerve crush. Cell Death Dis 4:67

94. Della Santina L, Ou Y (2017) Who's lost first? Susceptibility of retinal ganglion cell types in experimental glaucoma. Exp Eye Res 158:43–50

95. Duan X, Qiao M, Bei F, Kim IJ, He Z, Sanes JR (2015) Subtype-specific regeneration of retinal ganglion cells following axotomy: effects of osteopontin and mTOR signaling. Neuron 85:1244–1256

96. El-Danaf RN, Huberman AD (2015) Characteristic patterns of dendritic remodeling in early-stage glaucoma: evidence from genetically identified retinal ganglion cell types. J Neurosci 35:2329–2343

97. Majander A, Joao C, Rider AT, Henning GB, Votruba M, Moore AT, Yu-Wai-Man P, Stockman A (2017) The pattern of retinal ganglion cell loss in OPA1-related autosomal dominant optic atrophy inferred from temporal, spatial, and chromatic sensitivity losses. Invest Ophthalmol Vis Sci 58:502–516

98. Ou Y, Jo RE, Ullian EM, Wong RO, Della Santina L (2016) Selective vulnerability of specific retinal ganglion cell types and synapses after transient ocular hypertension. J Neurosci 36:9240–9252

99. Puyang Z, Gong HQ, He SG, Troy JB, Liu X, Liang PJ (2017) Different functional susceptibilities of mouse retinal ganglion cell subtypes to optic nerve crush injury. Exp Eye Res 162:97–103

100. Daniszewski M, Senabouth A, Nguyen QH, Crombie DE, Lukowski SW, Kulkarni T, Sluch VM, Jabbari JS, Chamling X, Zack DJ et al (2018) Single cell RNA sequencing of stem cell-derived retinal ganglion cells. Sci Data 5:180013

101. Doudna JA, Charpentier E (2014) Genome editing. The new frontier of genome engineering with CRISPR-Cas9. Science 346:1258096

102. Dominguez AA, Lim WA, Qi LS (2016) Beyond editing: repurposing CRISPR-Cas9 for precision genome regulation and interrogation. Nat Rev Mol Cell Biol 17:5–15

103. Konermann S, Brigham MD, Trevino AE, Joung J, Abudayyeh OO, Barcena C, Hsu PD, Habib N, Gootenberg JS, Nishimasu H et al (2015) Genome-scale transcriptional activation by an engineered CRISPR-Cas9 complex. Nature 517:583–588

104. Mandegar MA, Huebsch N, Frolov EB, Shin E, Truong A, Olvera MP, Chan AH, Miyaoka Y, Holmes K, Spencer CI et al (2016) CRISPR interference efficiently induces specific and reversible gene silencing in human iPSCs. Cell Stem Cell 18:541–553

105. Ratz M, Testa I, Hell SW, Jakobs S (2015) CRISPR/Cas9-mediated endogenous protein tagging for RESOLFT super-resolution microscopy of living human cells. Sci Rep 5:9592

106. Yang H, Wang H, Shivalila CS, Cheng AW, Shi L, Jaenisch R (2013) One-step generation of mice carrying reporter and conditional alleles by CRISPR/Cas-mediated genome engineering. Cell 154:1370–1379
107. Sander JD, Joung JK (2014) CRISPR-Cas systems for editing, regulating and targeting genomes. Nat Biotechnol 32:347–355
108. Stern JH, Temple S (2011) Stem cells for retinal replacement therapy. Neurotherapeutics 8:736–743
109. Wiley LA, Burnight ER, Songstad AE, Drack AV, Mullins RF, Stone EM, Tucker BA (2015) Patient-specific induced pluripotent stem cells (iPSCs) for the study and treatment of retinal degenerative diseases. Prog Retin Eye Res 44:15–35

Chapter 6
Surgical Approaches for Cell Therapeutics Delivery to the Retinal Pigment Epithelium and Retina

Boris Stanzel, Marius Ader, Zengping Liu, Juan Amaral,
Luis Ignacio Reyes Aguirre, Annekatrin Rickmann, Veluchamy A. Barathi,
Gavin S. W. Tan, Andrea Degreif, Sami Al-Nawaiseh, and Peter Szurman

Abstract Developing successful surgical strategies to deliver cell therapeutics to the back of the eye is an essential pillar to success for stem cell-based applications in blinding retinal diseases. Within this chapter, we have attempted to gather all key considerations during preclinical animal trials.

Guidance is provided for choices on animal models, options for immunosuppression, as well as anesthesia. Subsequently we cover surgical strategies for RPE graft delivery, both as suspension as well as in monolayers in small rodents, rabbits, pigs, and nonhuman primate. A detailed account is given in particular on animal variations in vitrectomy and subretinal surgery, which requires a considerable learning

B. Stanzel (✉)
Eye Clinic Sulzbach, Knappschaft Hospital, Sulzbach, Saar, Germany

Fraunhofer Institute for Biomedical Engineering, Sulzbach, Saar, Germany

Department of Ophthalmology, National University of Singapore, Singapore, Singapore
e-mail: boris.stanzel@kksaar.de

M. Ader · L. I. R. Aguirre
DFG Center for Regenerative Therapies Dresden (CRTD), Technische Universität Dresden, Dresden, Germany

Z. Liu
Department of Ophthalmology, National University of Singapore, Singapore, Singapore

J. Amaral
Stem Cell and Translational Research Unit, National Eye Institute, National Institutes of Health, Bethesda, MD, USA

A. Rickmann · S. Al-Nawaiseh · P. Szurman
Eye Clinic Sulzbach, Knappschaft Hospital, Sulzbach, Saar, Germany

V. A. Barathi · G. S. W. Tan
Singapore Eye Research Institute, Singapore, Singapore

A. Degreif
Fraunhofer Institute for Biomedical Engineering, Sulzbach, Saar, Germany

© Springer Nature Switzerland AG 2019
K. Bharti (ed.), *Pluripotent Stem Cells in Eye Disease Therapy*,
Advances in Experimental Medicine and Biology 1186,
https://doi.org/10.1007/978-3-030-28471-8_6

curve, when transiting from human to animal. In turn, however, many essential subretinal implantation techniques in large-eyed animals are directly transferrable to
human clinical trial protocols.

A dedicated subchapter on photoreceptor replacement provides insights on preparation of suspension as well as sheet grafts, to subsequently outline the basics of
subretinal delivery via both the transscleral and transvitreal route. In closing, a
future outlook on vision restoration through retinal cell-based therapeutics is
presented.

Keywords Age-related macular degeneration · Retinal pigment epithelium ·
Photoreceptor · Transplantation · Cell-based therapy · Cell replacement · Surgery ·
Anesthesia · Mouse · Rat · Rabbit · Pig · Nonhuman primate · Monkey · Preclinical
study · Vitrectomy · Immunosuppression

6.1   Introduction

Delivery of novel therapeutic agents to the retina, in particular for cell replacement
of the retinal pigment epithelium (RPE) and photoreceptors (PR), has gained considerable interest since the introduction of efficient protocols for the generation of
pluripotent stem cell (PSC)-derived cell transplants for clinical application. Initial
feasibility of RPE transplantation was demonstrated using primary cell sources by
Gouras et al. in monkeys [1] and subsequently also for photoreceptors in rats by
Silverman and Hughes [2]. Surgical clinical experience with replacement of dysfunctional or lost RPE in age-related macular degeneration (AMD) or photoreceptors in retinal dystrophies has been available since the 1990s [3, 4].

Since then, diverse distinct animal models as well as treatment modalities in
patients have been explored. While the RCS rat remains the gold standard for visual
testing for the FDA, athymic rats and humanized mice enable high-throughput testing, such as teratogenicity assays of human PSC-derived retinal cells as regards
human immune system characteristics. Rabbits and pigs enable refinement of surgical instrumentation and techniques, as well as immune suppression protocols immediately applicable to human use, with nonhuman primates offering a uniquely
precious opportunity to validate all prior findings in a foveate animal.

Given the complexity of photoreceptor transplantation a number of different
inherited mouse models characterized by photoreceptor dysfunction and degeneration have been introduced. These include autosomal recessive, dominant, and
X-linked models affecting either rods, cones, or both photoreceptor types. The different mouse models of retinal degeneration were derived either by spontaneous
mutations or genetic engineering. Phenotypically, the available mouse models often
resemble the disease phenotypes observed in humans. Particularly inherited retinal
degeneration can be quite precisely modeled by gene-modified mice, including
rapid to slow degeneration patterns spanning time lines from a few days to several

months for the loss of main parts of the ONL. However, proper mouse models for more complex retinal diseases like AMD have not been established. While specific aspects of AMD have been recapitulated in rodent models, the complete spectrum could not been reproduced, what might be not surprising given, for example, the short life span of nocturnal rodents besides the lack of a macula. Though mouse retinopathy models have been and further will be of utmost importance for developing photoreceptor transplantation strategies for the treatment of retinal degenerative diseases, the use of large, diurnal animal models is more and more recognized as an important translational step toward clinical application. Indeed, first pig models with inherited retinal dystrophies have been generated and given the recent advancements in gene editing using CRISPR/Cas technology more specific models are expected in the near future. Of interest may be further already available dog strains, that suffer from spontaneous mutations in vision related genes due to intensive inbreeding for the generation of particular strains [5]. Given the existence of a macula in nonhuman primates these animals represent highly reliable animal models. However, besides limitations in regard to organization, finances, time, and regulatory requirements for experimental work with monkeys, there are currently no genetic primate models of inherited retinal degeneration available. Though, first transplantations studies using human embryonic stem cell-derived retinal tissue were performed in photoreceptor degeneration monkey models, generated by acute chemical or light/laser damage [6].

Clinically, RPE delivery under the macula has been achieved with cell suspensions, RPE-choroid patches, isolated RPE sheets—both cultured and uncultured—and recently aided by a cell carrier as a cultured monolayer from human embryonic PSCs. Clinical subretinal neural retina transplantation techniques have utilized both suspensions [7] and isolated cadaveric fetal human retinal sheets [8].

Here we provide the required surgical essentials stratified by species and animal model to bring retinal derivatives of PSCs into the subretinal space in preclinical studies.

6.2 General and Anesthetic Considerations

6.2.1 Logistic and Management Advice

The animals should be held indoors in a specialized facility in an air-conditioned room with temperatures between 18 °C and 20 °C, exposure to regular daylight and in standardized individual cages with free access to food and water [9].

To ensure the animals' operative affinity, an animal health score sheet is followed. This includes the following definitive animal exclusion criteria: 20% weight loss compared to weight on admission; inability to eat or drink; behavioral abnormalities such as CNS signs, vocalization, hunched posture, shivering, decreased activity, immobility; apparent cyanosis of the animal; has cramps or cannot move in coordination; ataxia/paresthesia (e.g., paralyses); apathy; extreme automutilation (skin wounds, severed limbs).

All perioperative parameters should be carefully recorded and in case of abnormalities discussed with the surgeon and/or principal investigator in due time. Electronic record keeping and database filing is strongly encouraged. Animal IDs, such as tattoos, plastic tags, or even subdermal chips, should be considered, as they facilitate their identification during busy surgical days and follow-up examinations.

6.2.2 Animal Choices

During the last 2 decades, there has been extensive research with animal models of retinal diseases. To date, several species—including mice, rats, cats, dogs, rabbits, pigs, and nonhuman primates—have been used as models to provide valuable information on the cellular and molecular aspects of pathogenesis of retinal diseases.

6.2.2.1 Small Rodents

The Royal College of Surgeons (RCS) rat is widely used for research as a hereditary retinal dystrophies model. It was identified to be a mutation in Mertk gene and results in defective retinal pigment epithelium phagocytosis of photoreceptor outer segments [10]. The RCS rat remains an obligatory disease model to evaluate the efficiency of RPE cell therapy in preclinical regulatory studies accepted by the US food and drug administration (FDA) to approve clinical trials for RPE cell therapeutics. However, several studies provided evidence that diverse cell populations other than RPE transplanted into RCS rats have also beneficial effects on ONL preservation [11–14]. Therefore, detailed analysis has to be performed to identify the RPE-specific effects on photoreceptor rescue in this model. Additional small rodent models used for long term teratogenicity assays were also the athymic nude rats [15] and humanized mice [16].

There is a naturally occurring mouse model of (retinitis pigmentosa-like) retinal degeneration, called the rd mouse, which exists in several variants [17]. Due to the often rapid retinal degeneration, the transplantation may need to be done at a very early stage [18], which makes it more challenging to operate on half smaller eyes compared to RCS rats [19].

6.2.2.2 Rabbits

We recommend Dutch Belted rabbits weighing a minimum of 1.5 kg due to less fibrin reaction than Chinchilla Bastard Hybrid rabbits [9]. Another alternative are New Zealand Reds, which are a pigmented crossbreed with New Zealand Whites (albinos). The former have a very robust eye wall (rigid sclera) for retinal surgery and lesser bleeding tendency in the authors' experience. Albino rabbits (New Zealand White) are of limited value in our opinion, given aberrant RPE physiology and challenging contrasts of the retinal surface (due to the total lack of pigmentation), thus making subretinal manipulation difficult. One option to circumvent this challenge may be the use of chromovitrectomy dyes to stain the ILM such as triamcinolone or a commercially available mixture of trypan blue, brilliant blue G and PEG (Membrane Blue Dual® or Brilliant Peel Dual Dye ®).

6.2.2.3   Pigs

The structure of the retinal vascular system in pigs is very similar to that of humans, which makes them very useful for research on eye diseases especially for retinal diseases.

Domestic Pigs

It is recommended to utilize smaller size domestic pigs (body weight 8–12 kg) to avoid management issues. Domestic pig growth rate is very fast as compared to mini and miniature pigs. The disadvantages of domestic pig models include lack of availability of molecular reagents and antibodies, difficulty in maintenance, and less precise genetic characterization and manipulations.

Mini-Pig

To develop effective cell therapeutics for RPE/Retina, it is imperative to establish an animal model that is reproducible, closest in anatomy to the human eye and most representative of the human disease. There is an increased interest to use mini-pigs in ocular experimental studies due to their anatomical similarities with human eyes and as a substitute for nonhuman primates. Pharmacologically, the mini-pig eye behaves similarly to the human eye, making it ideal for testing new therapeutics and surgical approach.

Yucatan Pig

Recently, there is an increased interest to use Yucatan pigs in ocular experimental studies due to their anatomical and physiological similarities with human eyes and as a substitute for nonhuman primates. Pharmacologically and physiologically, the Yucatan pig eye behaves similarly to the human eye, making it ideal for testing new therapeutics.

At the NEI, we use Yucatan minipigs weighing 30–35 Kg (around 6–8 months old). Yucatan's are preferred because of their fundus pigmentation; allowing us to use a micropulse 532 nm laser 48 h before surgery to damage the RPE in the visual streak (cone rich area) inducing a retinal degeneration. The laser induced retinal degeneration created allowed us to evaluate post-surgical retinal recovery (OCT, mfERG) in transplanted areas.

Recently, Ross et al. described a pig model of Retinitis pigmentosa with a Rhodopsin mutation, [20], thus offering a potent large-eye animal model for photoreplacement studies.

6.2.2.4   Nonhuman Primates

Cynomolgus (*Macaca fascicularis*) and Rhesus monkeys (*Macaca mulatta*) weighing at least 3 kg and aged at least 2 years have been used by most investigators [21–25]. Some investigators used *Saimiri sciureus, a New World monkey* [26].

It appears as if rhesus macaques have a somewhat more robust foveal architecture for submacular surgery. Whenever feasible, the use of specific pathogen free (SPF) animals is recommended.

6.2.3   Immunosuppression

6.2.3.1   Mouse and Rat

Rodents were immunosuppressed with cyclosporine either by intramuscular injection (10 mg/kg) with cyclosporine blood level above 1500 µg/l [27] or when added (210 mg/l) to drinking water [28, 29], resulting in a mean cyclosporine blood level of 321 ± 21.9 µg/l [30]. Cyclosporine was given 1 day before transplantation until the end of the study. Lu et al. also included dexamethasone as an additional immunosuppressive drug by intraperitoneal injection (1.6 mg/kg/day) for 2 weeks after surgery [31]. Body weight and general physical condition of each animal were closely monitored every other day.

An alternative way is to use immunodeficient mouse disease model. Iraha et al. described two mouse models (NOG-rd1-2J and NOG-rd10) of end-stage retinal degeneration with immunodeficiency [32].

6.2.3.2   Rabbit

Immunosuppression regimens in rabbits have been utilized extensively for RPE replacement. It is recommended to induce the animals at least 3 days before starting the surgery to avoid immune sensation already at surgery.

The most simple is perhaps the use of 1–2 mg intravitreal preservative-free tri-amcinolone under short-term intramuscular general anesthesia and local numbing drops (e.g., Oxybuprocaine), if few weeks of experimental follow up are planned [33, 34]. We discourage the use of systemic dexamethasone for subretinal RPE xenograft protection, as this resulted in disruption and/or cell loss within scaffold-supported RPE monolayer grafts [33]. To avoid steroids, both local [35] and systemic cyclosporine [7] protocols are available.

A triple immunosuppression regimen (prednisone, cyclosporine, and azathio-prine) from the day of surgery throughout the experiment utilized by Del Priore [36] for aggregates of uncultured pig RPE sheets encapsulated in gelatin xenografted into the subretinal space of rabbits failed to provide protection at 4 and 12 week time points to both the graft and the host outer retina. This may be related to gelatin encapsulation being pro-inflammatory and was seen also by Stanzel et al. in human to rabbit RPE xenotransplantation experiments [33].

Our current recommendation is per oral Sirolimus (1 mg/day), Doxycycline (15 mg/kg/day) and Minocycline (15 mg/kg/day) started 3 days prior to surgery and continued throughout the experiment combined intravitreal triamcinolone at the time of surgery. The protocol is effective in protecting subretinal RPE xeno-grafts in animals with a sodium iodate compromised outer blood–retinal barrier, see Fig. 6.1.

An intravitreal injection of Sirolimus (220 µg, DE-109, Santen Incorporated) [37] with a documented efficacy of up to 2 months in nonvitrectomized animals may obviate the need for repeated intravitreal injections or laborious systemic

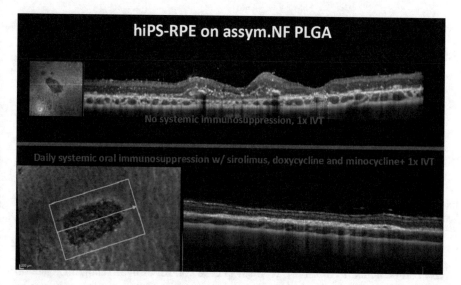

Fig. 6.1 Effect of quadruple immunosuppression on subretinal human iPS-RPE xenograft integration in rabbits with damaged blood–retinal barrier

application. Here the drug typically deposits to the anterior vitreous and can be readily recognized as a whitish precipitate. Thus if during vitrectomy extensive vitreous removal is avoided, which is also safer to avoid lens touches, then the slow release formulation can be preserved in place, and may not necessitate repeated injections.

6.2.3.3 Pig

Since scaffolds with human derived iPSC-RPE cells are introduced in the subretinal space, to minimize xenograft immune reaction, pigs are immunosuppressed starting 9 days before surgery and continued throughout until euthanasia. Tetracycline antibiotics doxycycline and minocycline are used orally with doses of 5 mg/Kg twice a day, mainly because of its suppressing effect on microglia activation. Steroids are used because of its broad-spectrum immunosuppressive effects. A loading dose of intramuscular methylprednisolone is used at doses of 5 mg/Kg, followed by similar daily oral single dose of prednisone. Both rapamycin (sirolimus) and tacrolimus are used to suppress adaptive immune response. Rapamycin is used orally with a loading dose of 2 mg, followed by a 1 mg daily dose. Tacrolimus is used in oral doses of 0.5 mg/day.

Koss et al. described pig model with perioperative sustained dexamethasone release implant (Ozurdex®, Allergan Inc.), along with a sophisticated continuous intravenous Tacrolimus application to protect monolayers of human embryonic derived RPE on porous parylene scaffolds, for further details, please refer to [38].

6.2.3.4 Monkey

Systemic immunosuppression drugs in monkeys include sirolimus, doxycycline and minocycline. The animal should be immunosuppressed 7–10 days prior and throughout the experiment to avoid immune sensation at the time of surgery and immune rejection after surgery.

Systemic sirolimus is given per oral as follows: On first day 2 tablets of 1 mg/tab sirolimus once a day (2 mg/day), thereafter 1 tablets of 1 mg/tab once a day (1 mg/day) until enucleation. Difficulties in administering sirolimus orally (monkeys are picking and throwing the drugs even when disguised with gel, pudding or marshmallow) can be overcome by crushing the tablets and mixing into chocolate milk or if unavoidable, animals under light sedation (ketamine 5 mg/kg BW, IM). Doxycycline and minocycline dosage is 15 mg/kg and divided into twice daily per oral application in all animals of both cohorts. As an alternative fall back strategy systemic prednisone 5 mg/kg and tacrolimus extended release (Astagraf) 0.5 mg/day can be added to the above regimen in case control of over xenograft rejection reactions is not adequate.

6.2.4 Anesthesia

6.2.4.1 Small Rodents (Mice and Rats)

Prior to starting anesthesia the animals are given food and water ad libitum and left in their own respective cage until needed. To ensure proper dosage of analgesics and anesthetics, the animal's weight is recorded. About 1–2 days prior to surgery pain management is started with Carprofen 5 mg/kg SC every 12–24 h or Metamizol 25 mg/drop in a solution 1:4, 0.2 to 2 drops every 6 h perorally.

Premedication is best achieved with a chamber for inhalation with isoflurane, alternatively an injection of Xylazine 5–10 mg/kg intraperitoneally can also be used. To avoid hypothermia you need a warm and soft pad throughout the procedure. Pupils are dilated by topical application of 1% tropicamide (1 drop) and a local anesthetic (to reduce dependence on general anesthesia), such as paraprocaine instilled on the ocular surface.

General anesthesia is then induced by intraperitoneal injection of a solution of ketamine 65 mg/kg (range 40–80 mg/kg), xylazine 13 mg/kg (range 10–15 mg/kg), and acepromazine 1.5–2 mg/kg; this will suffice for about 50 min. Observation of vital parameters, particularly hypothermia, is essential.

After the procedure the mouse/rat is taken to its own cage (isolation). The litter is covered with a tissue. Some food and possibly some gel cushions are placed on the bottom of the cage. The animal is closely observed until awake and has resumed eating and drinking. Thereafter regular follow up examinations of the operated eye, along with application of anti-inflammatory/antibiotic eye drops as per protocol are warranted.

6.2.4.2 Rabbit

General

Prior to anesthesia, the animal is allowed free access to food and water. Each rabbit is to be handled with care with good fixation of the rabbit, especially at the hind limbs since there is a high risk that they injure themselves. If the rabbit is agitated, it helps to place the palm of the hand over the ears (to pull them down to the side) and the extended index finger (of the same hand) onto the forehead, whilst gently pressing the head down. Once the rabbit has calmed down, it can be gently picked up by one hand sliding down to the scruff, whilst the other supports the bum (so the hind limbs do not hang free and start swinging/charging).

The animal's weight is recorded for anesthesia and analgesia dosage calculation. Pain management should be considered before, during, or directly after the surgery with, for example, meloxicam (2 mg/ml) 0.2 mg/kg SC every 24 h for 3–5 days.

The rabbit should be covered with a blanket to calm before the anesthesia injection. Premedication is achieved using intramuscular xylazine 0.2 mg/kg (IM) injection into the hind limb (gluteal muscle) and massage around injection site for 30 s. A subcutaneous infusion (at the scruff) of 10 ml/kg 0.9% saline is recommended to stabilize hydration before and after surgery. For induction of anesthesia use ketamine 35 mg/kg IM and xylazine 3 mg/kg, and butorphanol 0.1 mg/kg. The maintenance solution containing ketamine 35 mg/kg IM, xylazine 3 mg/kg, and butorphanol 0.1 mg/kg should be injected 0.2 ml/kg every 20 min or when the rabbit shows wake-up reflexes. Alternatively, consider placing an intravenous line into the (well heated) auricular vein, if IV anesthesia is considered.

The first sign of the anesthesia fading away is nystagmus and must be monitored by the surgeon. The first shot of IM anesthesia lasts about 30–60 min, depending on the rabbit's size, drug tolerance, fat layer, stress, and body temperature. Always confirm proper anesthesia, by verifying hypnosis, hyporeflexia, analgesia, and muscle relaxation of the animal. Ensure the proper body core temperature using rectal thermometer (normothermia 39 ± 1 °C). A heating pillow underneath the rabbit may often not be necessary or even dangerous, if the animal is properly wrapped in blanket. An oxygen line placed under the blanket cover supplied with a funnel wrapping around the snout significantly helps to avoid accidental hypoxic brain damage during anesthesia. If feasible intraoperatively, observation of heartbeat and breath frequency can further improve anesthetic success.

After the procedure, the rabbit should be carefully transferred into warm cage with fresh hay, consider either a blanket or heating lamp, and positioned as specified by the surgeon (operated eye facing up or down). Do not leave the animal unattended until it regains sufficient consciousness to maintain sternal recumbence. Regular observation is warranted until the rabbit resumes fluid and food intake.

Head Positioning

The optimal position of the rabbit under surgical microscope is with the nose slightly elevated through a mold of the blanket, so that it is level with ocular surface. Align the left eye perpendicular to microscope objective. We recommend using the left eye due to the positioning of the eye muscles. The eyelashes must be cut using scissors (some ointment on blade) to reduce postoperative infections.

Eye-Specific Preparations

It is recommended to dilate the pupil prior to anesthesia with 2.5% phenylephrine and 1% tropicamide 1 h before surgery every 10 min, repeated three times; this will yield a faster and more lasting mydriasis. To disinfect the eye use 2–3 drops of 0.1 g/ml povidone–iodine topically for 1 min and rinse with sterile BSS. Then, the eye should be covered with sterile drape with a precut opening in the middle for the eye and then covered with (sticky) surgical incision drape 12 × 17 cm. Methylcellulose or preferably a clinical grade dispersive viscoelastic lubricant should be added every 5–10 min to the operated eye, and lids in the nonoperated eye should be given lubricating ointment (e.g., Bepanthen®) and taped.

Postoperative eye care is recommended for 1 week twice a day, preferably with ointments containing a steroid and antibiotics (e.g., Isopto-Max®).

6.2.4.3 Pig

General

The last feeding should be about 12 h before the start of the surgery, while access to water is not restricted. The weight of the animal is recorded for the dosage of anesthesia and analgesia. For sedation and to enable transport the pig to surgical suite for intubation inject intramuscular 5.0 mg/kg telazol, 0.2 mg/kg butorphanol, and 0.035 mg/kg dexmedetomidine. For premedication, 0.02 mg atropine/kg deep IM is given. Then the monitoring probes are placed on the animal.

General anesthesia is induced by inhalation of 2–5% isofluorane. To prevent eye drift and possibly fatal oculocardiac reflexes during surgery application of neuromuscular block Rocuronium at 2 mg/kg IV; injection is repeated as necessary.

Analgesia is administered while pig is still anesthetized, 0.005–0.01 mg/kg Buprenorphine IM. Pigs are sedated during recovery phase with 1.0 mg/kg Diazepam IV and 1.1 mg/kg Acepromazine IM, to prevent eye trauma.

Head Positioning and Eye Specific Preparations

Pigs are position in the surgical table in lateral recumbency and the head position so the cornea is perpendicular to the microscope objective. Eye lashes are trim and povidone-iodine 10% is used to clean the skin around the eye. Povidone–iodine 5% should be used for the cul-de-sac.

A single-piece drape with an integral fluid collection is the most efficient method of draping. Drapes without an opening for the eye should be used and a cut is made in the drape after it is adhesively applied with the lids open. The flaps created are then folded over the lid margins and kept in place by the lid speculum. To better expose the surgical area, a temporal canthotomy is made and a traction suture to proptose the eye is applied to the inferior rectus; if needed the nictitans membrane can be retracted with sutures to better expose the nasal sclera. The corneal epithelium should be constantly irrigated with BSS or (a dispersive) viscoelastic.

6.2.4.4 Macaques/Nonhuman Primates

General

Presurgical preparation includes overnight fasting with no access to food from 6 pm onward to prevent regurgitation and vomiting during anesthesia and surgery.

In immediate preparation for surgery the monkey is sedated with intramuscular injection of Ketamine (10–20 mg/kg) and atropine 0.05 mg/kg (SC). An IM administration of Buprenorphine at 0.005–0.03 mg/kg BW (as pain reliever) is given 30 min to 1 h before surgery. The dose can be repeated 6 h after surgery, if necessary.

The animal should be weighed to ensure the accurate medication dosage. The 60–120 min procedure has to be performed under general anesthesia. A qualified veterinarian should perform intubation and general anesthetic procedure using 2% Isoflurane and an appropriately sized endotracheal tube according to the weight of the monkey. Vital signs such as electrical activity of the heart, respiratory rate, blood pressure, and oxygen saturation are tracked using monitoring equipment. The operation site is disinfected with alcohol, and finally swabbed with povidone iodine. Finally, the monkey is draped and ready for surgery.

During surgery, animals should receive vital signs monitoring including heart rate, respiration rate and/or end-tidal CO_2, SpO_2, temperature, and blood pressure.

Postprocedure analgesia is achieved with buprenorphine (0.005–0.1 mg/kg IM at 6- to 12-h intervals) or carprofen (2–4 mg/kg SC) for up to 3 days following all surgical procedures, and/or at any other time deemed necessary by the principal investigator or attending veterinary staff. The analgesia dosing depends on the animal's signs of pain (e.g., squinting, swollen eyelids, rubbing of the eyes, or tearing).

Animal Positioning

It is recommended to shave the periocular region to allow better access to the orbit and to improve adhesion of surgical draping. The optimal position of the monkey under surgical microscope is in supine position. To stabilize head position the use of a (custom-made) head mold is highly encouraged. Given occasional uncontrolled movements intraoperatively, we also suggest to secure the forehead with an adhesive tape which reaches to the surgical table. Ensure perfectly horizontal alignment of the limbal plane to ensure good surgical visualization and access to the macula. Retrobulbar injections to achieve a more proptosed position of the globe are very challenging.

Postoperatively, animals are then recovered in the recovery suite with the surgical eye stably positioned looking down for 1 h or longer to allow further reattachment of the macula.

Eye Specific Preparations

For pupil dilatation, the monkey should be given a mixture of 2.5% phenylephrine and 1% tropicamide eye drops prior to vitrectomy. To commence the procedure, tetracaine eye drops should be applied. Methylcellulose lubricant should be added every 5–10 min in the operated eye, and lids in the nonoperated eye should be taped. A sterile field is established by disinfection of periocular structures by exposure to 10% Povidone-Iodine and the ocular surface with 5% povidone–iodine for at least 1 min and rinse with sterile BSS. Thereafter, the animal has to be covered with a sterile drape and following placement of a lid speculum.

Postoperative application of Tobradex eye drops 5x/day and homatropine eye drops 3 times/day is followed for 1 week.

6.3   Subretinal RPE Graft Delivery

6.3.1   Mouse and Rat

6.3.1.1   Cell Suspension Preparation

Cultured RPE cells were trypsinized and resuspended (50,000 cells/μl) in BSS and loaded in a NanoFil syringe (RPE-kit, World Precision Instruments, USA) with 33-gauge bevelled needle. The needle was preflushed to avoid bubbles before injection.

6.3.1.2   Cell Sheet Preparation

An RPE monolayer was cultured on scaffolds (e.g., porous polyester terephthalate/PET) for 4–8 weeks. Before the transplantation, implants (monolayer on top of scaffolds) were cut off with a fine blade or puncher with the size approximate 0.7 mm × 1.2 mm for RCS rats [39, 40] or 0.5 mm wide section for mouse [32]. The implants were loaded in a custom-made implant tool [28, 40, 41] and/or kept in BSS before implantation. We recommend these steps to be carried out under a microscope station with a sterile surface in a clean room or surgical suite.

6.3.1.3   Animal Preparation Before Surgery

At the beginning of the surgery, the rats were placed on a sterile gauze under a surgical microscope. Eyes were dilated as described above. The skin around the eye was sterilized with 10% povidone–iodine and 1 drop of 5% povidone–iodine is applied to the eye surface followed 1 min later by a generous washout with BSS.

6.3.1.4 Cell Suspension Injection

After the conjunctiva was cut open by a 33-gauge needle, a full-thickness cut through sclera was made 1.0–1.5 mm posterior to the limbus at the temporal equator of the host eye [30, 42], 0.5–2 µl BSS was injected at same side of the sclera-choroid opening into the subretinal space [29, 30, 42] or through the diametrically opposed retina with a blunt needle gently touching the retina [43]. The bleb was further verified by putting a glass coverslip on the cornea to visualize the fundus under surgical microscope.

6.3.1.5 Cell Sheet Implantation

After the conjunctiva was opened, a small incision (approximately 0.8–1 mm for rats, 0.6–0.8 mm for mouse) was cut transsclerally approximately 1.5 mm posterior to the limbus at the temporal equator of the host eye by a 27 gauge needle until the choroid is exposed. A transcorneal anterior chamber paracentesis is performed by a 31-gauge needle to reduce IOP. A local retinal bleb is created by injecting approximately 5 µl BSS through the transscleral opening into the subretinal space by a 33-gauge or smaller blunt needle with a NanoFil syringe (RPE-kit, World Precision Instruments, USA). The bleb is further verified by putting a glass coverslip on the cornea to visualize the fundus under surgical microscope. The choroid was cut with a fine blade, while the retina remains intact. Occasional bleeding can be stopped with a cotton bud. The transplant is gently placed into the area of the subretinal bleb through the choroidal incision using a custom-made implantation tool [28, 40, 41] or 25 G intraocular forceps (with rigid scaffolds, such as PET). Optical coherence tomography (OCT) can be used to confirm the position of the implant immediately following the surgery. At the end of the surgery, the incision is closed with 10–0 Vicryl sutures.

6.3.2 Rabbit Surgery

6.3.2.1 Instrument Preparation

Establish and maintain a sterile field, by working in a closed room, wearing surgical scrubs, facial mask and hair cover. Disinfect hands prior wearing sterile surgical gloves. Place the sterilized instruments on a sterile drape. Place 1 ml syringe filled with 40 mg/ml triamcinolone attached to a 27 G needle for injection, 10 ml syringe with Balance salt solution (BSS), and 5 ml syringe of lubricant on drape. Also, place 3-0 silk, 7-0 vicryl, ocular sticks (to stop conjunctival/scleral bleeding), twister gauze sponges, wound closure strips (to fixate the vitrectomy tip tubing), and chandelier endoillumination fiber wire on a drape [9, 44]. The 27 G chandelier endoilluminator has to be connected to a light machine. Connect a vitrectomy set including high speed vitrector and (Venturi) cassette to a vitrectomy machine. A 500 ml oph-thalmic grade BSS bottle has to be connected to the cassette according to manufac-turer's instructions. Some vitrectomy machines will require priming of the system to be operational.

6.3.2.2   Vitrectomy

First, make a lateral canthotomy and proptose and secure the eye with 3–0 silk using inverted caliper. Then, perform a conjunctival peritomy and incise the conjunctiva with Vannas' scissors close to the limbus, but far enough from the blood vessels (~1 mm distance). Dissect the conjunctiva by creating a "T-cut" by enlarging the peritomy with the scissor parallel to the limbus and then incise the conjunctiva verti-cally in form of a "T" for about 6–7 mm. Carefully separate conjunctiva/tenon by blunt dissection. A two-port, 25 G core vitrectomy can be performed through a non-contact, wide-angle system with a 27 G chandelier illumination system.

To create a sclerotomy, use a 25 G microvitreoretinal (MVR) blade or 25 G flat head trocar at 4 o'clock on left eye (OS) is carefully inserting the sharp tip of the blade in the direction toward the optic nerve. We recommend starting 3.5–4 mm behind the limbus. This will result in a transretinal rather than trans pars plana approach, but is thereby lens-sparing. Then, slowly retract the blade in the same direction and avoid enlarging the sclerotomy. Insert the custom side port-infusion cannula [45] and suture it using a 7-0 Vicryl suture and set the intraocular pressure (IOP) at 15–24 mmHg (at lower range if valved trocars with good seal are avail-able). The second sclerotomy should be performed with a 27 G needle at 10 o'clock on OS. Then, insert a 27 G chandelier light into flat head trocar and fixate it with sticky tape and turn on the light source at ca. 20–30% (as per surgeon preference). To better visualize the ocular fundus, the corneal epithelium typically requires scraping with a surgical blade (judge by appearance of fine intraepithelial bullae).

Similar to the first sclerotomy, perform a sclerotomy at 2 o'clock on OS and (pre) place u-shaped 7-0 sutures around the sclerotomy without tying the knot if no tro-cars are used, otherwise secure the trocar with releasable sutures. Then insert the vitrectomy cutter tip by strictly pointing toward the posterior pole.

Start the vitrectomy around the entry port, to relieve vitreous traction at the trans-retinal entry site (see above) and ensure infusion flow. Meanwhile, 20 units/ml hep-arin and 0.001 mg/ml epinephrine should be added into the BSS infusion solution in parallel to reduce fibrin reaction. As heparin and epinephrine are not injected intra-ocularly, their effects are delayed depending on the infusion flow rate.

Then continue over the optic disc and the fibrae medullares using high speed vitrector by cutting the vitreous gel into small pieces at max 2000–8000 cuts/min and aspirating at max 100–200 mmHg [46]. A posterior vitreous detachment (PVD) has to be performed by separating the vitreous humor from the retina by holding the cutter over the posterior pole and (if feasible gently) superior of the optic disc, while aspirating only at max 50–200 mmHg without cutting (depending on how well sealed the system is, too much vacuum if eye wall collapse). The posterior hyaloid face can be engaged in aspiration mode with the cutter, immediately inferior to the optic disc and then separated from the inner limiting membrane (ILM) using mechanical traction [9]. To visualize and facilitate (near total) removal of the floating vitreous over the posterior pole and midperiphery during vitrectomy inject ca. 50 μl (40 mg/ml) triamcinolone or diluted fluorescein (ca. 0.1 mg/ml) intravitreally and then thoroughly remove it with vitrectomy over the entire posterior pole. Avoid crossing over under the lens; if more peripheral removal is necessary we suggest using a third well sealing valved 25 G or 27 G trocar. Indentation to shave the peripheral vitreous (optionally by a skilled assistant) is recommended if gas or oil tamponade is desired; we recommend performing this step after the actual implantation procedure in an air-filled eye as the eye wall will become very unstable.

6.3.2.3   Loading Implantation Instrument (Shooter)

Herein we describe the preparation of an instrument for subretinal implantation of RPE monolayer grafts on cell carriers; for subretinal injection of cell suspensions a commercially available 38 G Teflon tip cannula (MedOne) can be utilized. An air bubble to initiate the bleb retinal detachment and then, immediately following subretinal deposition of the RPE cell suspensions, another air bubble to seal the retinotomy, can facilitate the procedure [47]. For more details on loading the cannula with a suspension, the interested reader is referred to [8].

An RPE cell culture, regardless of source, should be rinsed prior to preparation of the implant three times with calcium- and magnesium-containing Hank's balanced salt solution (CM-HBSS).

A standard cell culture dish (100 × 20 mm) with 10 ml ophthalmic grade BSS should be centered under a light microscope. With a sharp, oval, custom-made hollow needle a 2.4 × 1.1 mm implant can be punched out to obtain a flat, bean-shaped substrate with two long edges and two round edges [9]. Then, gently flood the needle through the second port with BSS to flush out the implant. Optionally cut one round end of implant (<0.5 mm), just to obtain a third edge. Ensure that the implant is in the right orientation with the monolayer upside on the cell carrier. To change positioning carefully use two scalpels. Then, push the implant gently and completely into the shooter instrument using a needle holder until all of the implant is secured inside of the tip. The plunger should remain retracted and the "loaded" shooter tip should be kept under CM-BSS in dark until the moment of implantation.

6.3.2.4   Implantation

Approach the neural retina with an extendible 38 G or 41 G subretinal injection needle connected to a gastight syringe, ensuring that all air bubbles have been evacuated from tubing. Then, create a bleb retinal detachment (bRD) (approximately 2–3 disc diameters) by slowly injecting 20–30 μl BSS subretinally by an assistant, with the IOP set to 25 mmHg or less. Two bRD per eye can be raised safely. The same aforementioned procedure suffices to deliver a suspension of RPE, see above and [8].

A retinotomy enlarged to ca. 1.5 mm with a vertical 25 G VR-scissors is then created at the base of the 20–30 μl, so that the subretinal space is accessible for implantation or further maneuvering. The RPE underneath the bRD can be atraumatically scraped at high IOP (30–50 mm Hg), with a custom made 20 G extensible Prolene-loop instrument [44].

Before implantation, the sclerotomy at 4 o'clock (OS) must be enlarged (precisely) with a 1.4 mm incision knife to 20 G approach. Attempt passing through the sclerotomy with a 20 G shooter dummy and enlarge as needed to ensure smooth, yet snug transition of the loaded shooter. Then, pass with the loaded shooter through the sclerotomy at 15–25 mmHg [9]. Approach the retinotomy edge and eject the implant subretinally from an epiretinal position. The implant may be adjusted with half-closed 23 G scissors, forceps, or 38 G needle to make sure it is positioned well under the retina, away from the retinotomy. The RPE monolayer transplants should be placed cell-carrier-side down on bare Bruch's membrane or intact host RPE, ensuring that the xenografted RPE face photoreceptors. Drainage of subretinal fluid is optional, as the bRD will spontaneously resolve within few days or less [48]. If desired a slow-paced fluid-air exchange (FAX) with brush-tip silicone active extrusion cannula, rather than using perfluorocarbon liquids (PFC) based bRD flattening followed by FAX is preferable to avoid subretinal PFC entrapment. We discourage the use of expandable gases.

6.3.2.5 Wound Closure

After removing 27 G chandelier and 25 G infusion cannula, suture all sclerotomies. Prior to suturing the last sclerotomy, inject 25–50 μl (40 mg/ml) triamcinolone intravitreally. The IOP should be checked by palpation and adjusted by injection of BSS via 30 G needle/syringe, if needed. Then, suture the conjunctiva with 7-0 vicryl and remove the proptosing 3-0 silk sling slowly. Avoid the deep orbital venous plexus as this may lead to uncontrollable hemorrhage. Concluding, suture the temporal canthotomy with 5-0 silk and add dexamethasone/antibiotic ointment under the lid.

6.3.3 Pig Surgery

6.3.3.1 Instrument and Equipment Preparation

Instrumentation and equipment depends on the surgeon's preferences. We use an Alcon Constellation® vitrectomy system and a Zeiss OPMI Lumera 700 microscope equipped with the Resight® noncontact fundus viewing system and Rescan 700 for intraoperative OCT (iOCT). The microscope is connected to the Alcon NGENUITY®

3D Visualization System for heads-up 3D visualization of surgical maneuvers. Hand instruments are best kept on a Mayo stand between the surgeon and the assistant with a back table placed behind for other surgical tools and disposables. A Mayo stand over the animal carries the vitrectomy probe, endolaser, bipolar handpiece, infusion and extrusion tubing. Powder free gloves should always be used.

6.3.3.2 Vitrectomy and Induction of Retinal Detachment

A four-port pars plana vitrectomy is done 3.5 mm from the limbus. Only the nasal port conjunctiva is dissected to allow widening of the sclerotomy during implantation. Transconjunctival 25 G valved entry systems are used to ensure working in a closed system. To ensure sutureless ports, blades are introduced using a single plane 15° incision to create a scleral tunnel; only the nasal port blade is introduced perpendicular to the sclera with the flat edge in horizontal position to ensure a linear incision when widening the sclerotomy. The infusion port is located at the level of the external canthus and the infusion pressure set at 45 mmHg. The endoilluminator (chandelier) is located inferior nasally and 2 working ports (superior nasal and superior temporal) are made. A central and midperipheral vitrectomy is done using maximum cutting speed and aspiration (7500–10,000 cpm/650 mmHg). The lens should be spared so anterior vitreous cortex vitrectomy is avoided. A posterior vitreous detachment is induced using maximum aspiration beginning in the border of the optic nerve; if difficult, preservative-free triamcinolone acetonide is used to facilitate posterior vitreous identification. Triamcinolone is aspirated in the vitrectomy tip and released using the proportional reflux mode function in the foot pedal. A localized retinal detachment is produced in the laser-induced retinal degeneration area (visual streak) using an extendable PolyTip® 25/38 G cannula connected through a MicroDose™ injection kit (MedOne Surgical) to the viscous fluid injector system (VFI) set at 18 mmHg allowing precise food pedal control of the injected BSS. A 2.5 mm retinotomy is made with curved scissors. The temporal location of the infusion port prevents collapse of the retinal detachment due to turbulence when the nasal port is widened using a 2.4 mm slit knife (Mani®). We use a custom-made tissue clamp to close the enlarged sclerotomy while preparing for implantation.

6.3.3.3 Implantation

Custom made injector design consists of hydraulically operated plastic ergonomic handle with an S-shape metal cannula and flattened plastic tip attached through the MicroDose™ injection kit to the VFI system where viscous fluid is replaced by a hyaluronic acid aqueous solution allowing for foot pedal control atraumatic delivery of implants into the subretinal space, see Fig. 6.2. This instrument accommodates a 2 × 4 mm oval shaped native tissue-like implant (iPSC-RPE monolayer on PLGA scaffold) that is cut from the transwell plate with a custom-made trephine. Transplanted cells are loaded in the injector manually using a syringe with hyaluronic

Fig. 6.2 Subretinal implantation of a PLGA-scaffold supported iPS-RPE monolayer in pig. (**a**) Foot-pedal control subretinal release of iPSC-RPE PLGA patch from injector in micropulse induced retinal degeneration area (visual streak). (**b**) iOCT visualization of subretinal iPSC-RPE PLGA patch. Subretinal fluid is aspirated with extrusion cannula to flatten the retina while retinotomy is closed by apposition

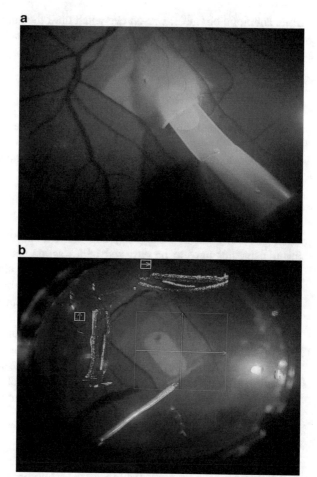

acid aqueous solution connected to the MicroDose™ injection kit through a two-way valve [49].

6.3.3.4   Ending Surgery

After implantation, the injector is released and the nasal sclerotomy temporally closed with the custom-made tissue clamp while fluid air exchange is used to reattached the bleb retinal detachment. The retinotomy is closed by apposition without retinopexy while iOCT confirms retinal flattening and implant location, see Fig. 6.2. The nasal sclerotomy is closed under air (nylon 8-0), air/gas exchange is made and ports withdraw. Always withdraw the infusion port at the end. Nasal conjunctiva is suture with Vicryl 7-0 and canthotomy with Vicryl 5-0. Administer subconjunctival 0.1 ml Depo-Medrol (20 mg/ml), 0.4 ml Gentamicin (100 mg/ml) followed with a triple antibiotic ointment. Use an eye patch until the animal recovery. Ketoprofen 3 mg/Kg is used for 3 days and triple antibiotic ointment is applied BID for 5 days.

6.3.4 Monkey Surgery

6.3.4.1 Instrument and Machine Preparations

Are identical to the general recommendations given for rabbits in Sect. 6.3.2.

6.3.4.2 Vitrectomy

First, perform a lateral canthotomy and insert a speculum. If globe exposure is insufficient, lids have to be retracted with traction sutures. Alternatively, a 360° conjunctival peritomy could be performed and a 3/0 silk sling suture is passed underneath rectus muscles to allow further exposure by suitable rotation of the globe. Preselecting (older) animals with good globe exposure can further facilitate surgical parameters.

Perform a conjunctival peritomy and incise the conjunctiva with a Vannas scissor close to the limbus, but far enough from the blood vessels. Dissect the conjunctiva by creating a "T-cut" by enlarging the peritomy with the scissor parallel to the limbus and then incise the conjunctiva vertically in form of a "T" for about 6–7 mm. The conjunctiva and Tenon are carefully separated from sclera by blunt dissection by the opening Vannas' scissor branches.

To perform a sclerotomy, use a 25 G microvitreoretinal (MVR) blade or 25 G trocar at 8 o'clock on right eye (OD) by carefully inserting the blade at 3 mm behind the limbus in direction toward the center of the globe. Then, slowly retract the blade in the same direction and avoid enlarging the sclerotomy. Insert a custom side port-infusion cannula and suture it using a 7-0 Vicryl suture and set the intraocular pressure (IOP) at 20 mmHg [45]. For a four-port vitrectomy, sclerotomies should be performed with a 25 G trocar at 2, 4, 8, and 10 o'clock on OD and preplace u-shaped 7-0 sutures around the sclerotomy with a releasable knot. Illumination is preferable with handheld endoillumination, rather than a trocar-based chandelier light source for improved maneuverability and light levels. The cornea should be lubricated and meticulously protected throughout the procedure with a mixture of ophthalmic grade viscoelastic (e.g., Viscoat®) and BSS; in case of corneal epithelial edema an abrasion with a surgical blade should not be deferred to improve fundus visualization (and surgical procedure time). Noncontact wide-angled fundus lenses attached to a surgical microscope (optionally equipped with an intraoperative OCT) are preferable for the rather small and often deep-set eyes.

Begin with core vitrectomy and vitreous removal around the main instrument ports. To visualize the vitreous fibers and facilitate surgical induction of posterior vitreous detachment inject ca. 20–50 µl (40 mg/ml) triamcinolone or diluted

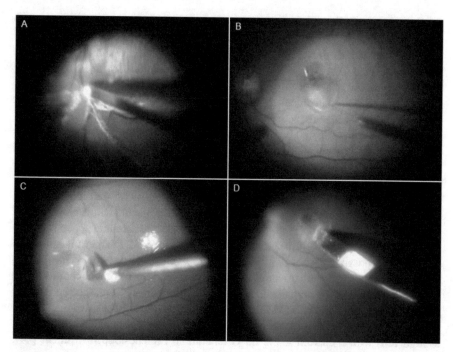

Fig. 6.3 Submacular hES-RPE Implantation sequence in nonhuman primate. (**a**) Shows triamcinolone-aided detachment of posterior cortical vitreous from retinal surface, (**b**) shows subretinal BSS injection to induce detachment of the macula, (**c**) shows retinotomy temporal and distal to the immediate macular arteriole, (**d**) submacular implantation of hES-RPE monolayer on a porous polyester scaffold

fluorescein (ca. 0.1 mg/ml) intravitreally. The posterior vitreous detachment (PVD) is performed by separating the vitreous gel from the retina by holding the vitrector over the optic disc at maximum vacuum without cutting, see Fig. 6.3a. Avoid elevating IOP to values beyond 30 mmHg as this may result in inner retinal damage. If engagement of the posterior hyaloid face cannot be achieved with the vitrectomy cutter alone, then use an end-gripping forceps to create a tiny break in the cortical vitreous near the optic disc and repeat. Upon creating the PVD, remove the vitreous skirt up to the vitreous base.

6.3.4.3   Preparing Implantation Instrumentation

Following same recommendation as in the rabbit section for scaffold supported RPE monolayer grafts. An alternative delivery technique for monolayer supported by its basement membrane was in part detailed (and tested in rabbits) by Kamao et al. [24, 50]. For RPE suspension grafts the recommendations are similar to what was given in the rabbit sect. 6.3.2.3, further details can be obtained by the interested reader in [22, 24, 51].

6.3.4.4 Implantation

Approach and blanch the neural retina with an extendible 38 G subretinal injection needle connected to a gastight syringe (e.g., 100 µl Hamilton) with BSS, ensuring that all air bubbles have been evacuated from tubing. Then, create a gentle controlled bleb retinal detachment (bRD) by an assistant of about 2–3 disc diameter (DD) involving the macula with subretinal injection of BSS, see Fig. 6.3b.

Wilson et al. investigated the use instrumentation and route for subretinal injection of RPE cell suspensions in rhesus macaques. Following a trans-pars plana induction of a bRD, a 30 G cannula reproducibly delivered a suspension graft in the subretinal space via a trans-scleral approach [51].

For scaffold supported RPE monolayer grafts, a vertical 1.5 mm long temporal retinotomy is created distal to a temporal arteriole of the macula with a vertical 25 G VR-scissors at elevated IOP (ca. 30 mm Hg), so that the subretinal space is accessible for implantation or further maneuvering, see Fig. 6.3c. Diathermia should to be avoided if possible.

Before implantation, the trocar at the main right hand sclerotomy is removed, 7/0 vicryl sutures need to be preplaced, and the sclerotomy must be enlarged (precisely) with a 1.4 mm incision knife (MVR blade) to fit subsequent 20 G instrumentation.

The submacular RPE removal underneath the bRD can be achieved by scraping at high IOP with a custom made 20 G extendible Prolene-loop instrument [44]. Attempt passing through the sclerotomy with a 20 G shooter dummy and enlarge as needed to ensure smooth, yet snug transition of the loaded shooter. Meanwhile, the custom subretinal implant shooter need to be loaded by a sterile assistant as specified above. The graft loaded into the shooter instrument is then handed to the surgeon, who will pass it swiftly through the sclerotomy ideally at 20 mmHg to then eject the implant from epiretinal into the subretinal space. Upon removal of the shooter instrument, immediately close the main sclerotomy by preplaced 7-0 vicryl sutures to avoid ocular hypotension. Alternative approaches have been described for RPE monolayer grafts by Kamao et al. [24, 50].

The implant is preferably adjusted with the 38 G Teflon cannula (as it carries the least risk for Bruch's membrane injury) to make sure it is positioned well under the macula, away from the retinotomy. The RPE monolayer transplants should be placed cell-carrier-side down on bare Bruch's membrane with the xenografted RPE facing photoreceptors.

Fluid air exchange is performed via active extrusion through the left hand trocar (or newly inserted right hand trocar) using a brushed silicone soft tip cannula. Gentle subretinal fluid aspiration from the bleb retinal detachment and retinotomy edge apposition is attempted at very low vacuum settings. Laser retinopexy to stabilize the retinotomy edges should be avoided. All steps of the submacular implantation may be monitored or guided if possible with intraoperative optical coherence tomography (iOCT). The left hand trocar port can be then used to for 1–2 mg preservative free triamcinolone (e.g., Triesence®) instillation.

6.3.4.5   Wound Closure

After removing 25 G chandelier and 25 G infusion cannula, remove all trocars and suture the sclerotomies. The IOP should be checked by palpation and adjusted if needed. Suture the conjunctiva with 7-0 vicryl and the temporal canthotomy with 5-0 silk. Polymyxin/neomycin/bacitracin antibiotic ointment is administered into the conjunctival sac.

6.4   Photoreceptor Transplantation

6.4.1   Considerations for Photoreceptor Graft Preparation

Retinal sheet transplantation offers the possibility of targeting both photoreceptors and RPE degeneration, and it can also consistently achieve a large continuous layer of photoreceptors with outer segments in contact with host or donor RPE, in recipients with a severely degenerated photoreceptor layer [52]. Due to the availability of stratified retinal tissue generated from pluripotent stem cell-derived retinal organoids, first studies provide evidence for proper orientation and polarization of in vitro-generated donor transplants including RPE contacts and synapse formation to endogenous second order neurons resulting in some functional improvements in the blind rd1 mouse model [53]. However, sheet transplantation approaches also show several limitations, when compared with the injection of suspended cells. The procedure is difficult and requires extensive training, as well as the acquisition of custom devices to deliver the sheet. The isolation of pure photoreceptor sheets from fetal or retinal organoid sources remains challenging, as in most cases inner retinal cell populations, that is, bipolar, horizontal, or amacrine cells besides Müller glia, are still attached to the photoreceptors, thus interfering with proper graft–host connectivity. Furthermore, transplantation of sheets frequently leads to rosette formation within the photoreceptor layer thus disrupting photoreceptor—RPE interactions, resulting in limited outer segment phagocytosis and consequently reduced chromophore recycling [53].

In contrast, the transplantation of dissociated cells, despite also requiring extensive training to master the subretinal injection procedure, has the advantage that only a small, self-healing incision is necessary for cell delivery, thus reducing tissue trauma. A minimal invasive procedure will be of particular importance in patients with advanced retinal degeneration, as such disease stages are associated with significant tissue thinning, which increases potential complications, such as retinal ruptures or complete retinal detachment. While preclinical studies provided evidence for proper maturation of donor photoreceptors, including synapse and outer segment formation following suspension injections, donor suspension transplants mainly generate disorganized cell clusters in the subretinal space, lacking a strict apical-basal orientation [54, 55].

6.4.2 Photoreceptor Transplant Delivery Strategies

The delivering technique of donor photoreceptors is another factor that significantly impacts the efficiency of transplantation. Subretinal transplantation in rodents can be performed by two alternative routes: transscleral and transvitreal injection, see Figs. 6.4 and 6.5.

While the trans-scleral approach allows relatively easy accessibility to the subretinal space (Fig. 6.4), visibility during the injection is reduced. If the instrument nozzle is placed at the wrong angle, it can damage Bruch's membrane, disrupting the RPE and the blood–retinal barrier (BRB), which in turn may cause bleeding and/or infiltration of the transplant with lymphocytes and macrophages, thus leading to transplant loss or rejection.

Transvitreal injections offer a clear visualization of the injection site and of the blood vessel network, thus reducing the risk of hemorrhage (Fig. 6.5). However, it induces some local retinal gliosis when the needle pierces through the retina.

Two methods for cell delivery in the subretinal space have been proposed: the transplantation of whole retinal sheets and the injection of suspended disaggregated cells. Retinal sheet transplantation methods were first developed in the 1990s for the delivery of gelatin embedded donor tissue, which is rolled up to fit into a round nozzle, and then unfolded after insertion into the subretinal space [56]. This method requires the formation of a subretinal bleb, as well as significant amounts of fluid to deliver the sheet, which causes trauma to both host and donor tissue. Retinal reattachment after delivery is also necessary. An improvement over this method was proposed later, in which a custom implantation device allowed a gentle placement of the retinal sheet, in the correct orientation and with minimal amount of fluid [57]. This variation does not require retinal reattachment and has been shown to produce cell integration within a degenerating retina [58].

6.4.3 Future Outlook for Photoreceptor Replacement

While photoreceptor transplantation is currently still at the preclinical stage, it is expected that, with the acquired knowledge in subretinal delivery of stem cell-derived RPE in first patients [59–62], clinical transplantation of photoreceptors will be performed within the coming years. Improved phenotyping of the degenerating retina with defining specific disease stages will be of utmost importance to identify the proper surgical method for donor cell delivery. Besides the degree of photoreceptor loss, additional environmental factors within the retina, like glial reactivity and scar formation, RPE constitution, hemorrhage, and inflammation might have significant influence on transplantation success.

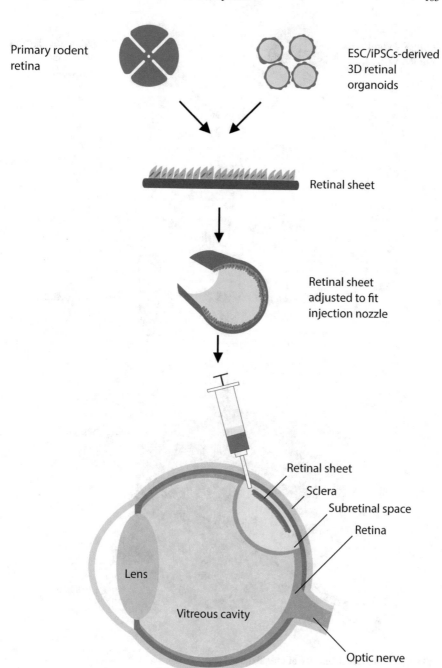

Fig. 6.4 Transscleral delivery of retinal transplants

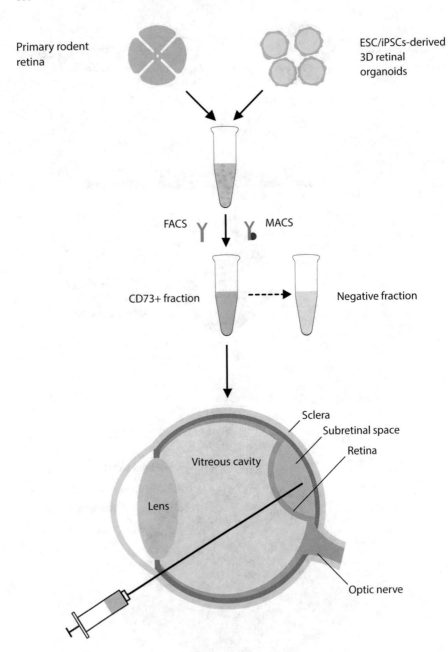

Fig. 6.5 Transvitreal delivery of retinal transplants

References

1. Gouras P, Flood MT, Kjeldbye H (1984) Transplantation of cultured human retinal cells to monkey retina. An Acad Bras Cienc 56(4):431–443
2. Silverman MS, Hughes SE (1989) Transplantation of photoreceptors to light-damaged retina. Invest Ophthalmol Vis Sci 30(8):1684–1690
3. Machemer R, Steinhorst UH (1993) Retinal separation, retinotomy, and macular relocation: II. A surgical approach for age-related macular degeneration? Graefes Arch Clin Exp Ophthalmol 231(11):635–641
4. Das T, del Cerro M, Jalali S, Rao VS, Gullapalli VK, Little C et al (1999) The transplantation of human fetal neuroretinal cells in advanced retinitis pigmentosa patients: results of a long-term safety study. Exp Neurol 157(1):58–68
5. Miyadera K (2014) Inherited retinal diseases in dogs: advances in gene/mutation discovery. Dobutsu Iden Ikushu Kenkyu 42(2):79–89
6. Shirai H, Mandai M, Matsushita K, Kuwahara A, Yonemura S, Nakano T et al (2016) Transplantation of human embryonic stem cell-derived retinal tissue in two primate models of retinal degeneration. Proc Natl Acad Sci U S A 113(1):E81–E90
7. Crafoord S, Algvere PV, Kopp ED, Seregard S (2000) Cyclosporine treatment of RPE allografts in the rabbit subretinal space. Acta Ophthalmol Scand 78(2):122–129
8. Petrus-Reurer S, Bartuma H, Aronsson M, Westman S, Lanner F, Kvanta A (2018) Subretinal transplantation of human embryonic stem cell derived-retinal pigment epithelial cells into a large-eyed model of geographic atrophy. J Vis Exp 131. https://doi.org/10.3791/56702
9. Al-Nawaiseh S, Thieltges F, Liu Z, Strack C, Brinken R, Braun N et al (2016) A step by step protocol for subretinal surgery in rabbits. J Vis Exp 115:53927
10. D'Cruz PM, Yasumura D, Weir J, Matthes MT, Abderrahim H, LaVail MM et al (2000) Mutation of the receptor tyrosine kinase gene Mertk in the retinal dystrophic RCS rat. Hum Mol Genet 9(4):645–651
11. Lawrence JM, Sauve Y, Keegan DJ, Coffey PJ, Hetherington L, Girman S et al (2000) Schwann cell grafting into the retina of the dystrophic RCS rat limits functional deterioration. Royal College of surgeons. Invest Ophthalmol Vis Sci 41(2):518–528
12. Inoue Y, Iriyama A, Ueno S, Takahashi H, Kondo M, Tamaki Y et al (2007) Subretinal transplantation of bone marrow mesenchymal stem cells delays retinal degeneration in the RCS rat model of retinal degeneration. Exp Eye Res 85(2):234–241
13. Wang S, Girman S, Lu B, Bischoff N, Holmes T, Shearer R et al (2008) Long-term vision rescue by human neural progenitors in a rat model of photoreceptor degeneration. Invest Ophthalmol Vis Sci 49(7):3201–3206
14. Francis PJ, Wang S, Zhang Y, Brown A, Hwang T, McFarland TJ et al (2009) Subretinal transplantation of forebrain progenitor cells in nonhuman primates: survival and intact retinal function. Invest Ophthalmol Vis Sci 50(7):3425–3431
15. Diniz B, Thomas P, Thomas B, Ribeiro R, Hu Y, Brant R et al (2013) Subretinal implantation of retinal pigment epithelial cells derived from human embryonic stem cells: improved survival when implanted as a monolayer. Invest Ophthalmol Vis Sci 54(7):5087–5096
16. Zhao T, Zhang ZN, Westenskow PD, Todorova D, Hu Z, Lin T et al (2015) Humanized mice reveal differential immunogenicity of cells derived from autologous induced pluripotent stem cells. Cell Stem Cell 17(3):353–359
17. Chang B, Hawes NL, Hurd RE, Davisson MT, Nusinowitz S, Heckenlively JR (2002) Retinal degeneration mutants in the mouse. Vis Res 42(4):517–525
18. Aramant RB, Seiler MJ (2002) Retinal transplantation–advantages of intact fetal sheets. Prog Retin Eye Res 21(1):57–73
19. Remtulla S, Hallett PE (1985) A schematic eye for the mouse, and comparisons with the rat. Vis Res 25(1):21–31

20. Ross JW, Fernandez de Castro JP, Zhao J, Samuel M, Walters E, Rios C et al (2012) Generation of an inbred miniature pig model of retinitis pigmentosa. Invest Ophthalmol Vis Sci 53(1):501–507
21. Stanzel BV, Amaral J, Maminishkis A, Liu Z, Ilmarinen T, Hongisto H et al (2017) Seeing the invisible with intraoperative OCT in surgical vitreoretinal animal research for upcoming clinical applications. Invest Ophthalmol Vis Sci 58:3389
22. Sugita S, Iwasaki Y, Makabe K, Kamao H, Mandai M, Shiina T et al (2016) Successful transplantation of retinal pigment epithelial cells from MHC homozygote iPSCs in MHC-matched models. Stem Cell Reports. 7(4):635–648
23. Sugita S, Iwasaki Y, Makabe K, Kimura T, Futagami T, Suegami S et al (2016) Lack of T cell response to iPSC-derived retinal pigment epithelial cells from HLA homozygous donors. Stem Cell Reports. 7(4):619–634
24. Kamao H, Mandai M, Okamoto S, Sakai N, Suga A, Sugita S et al (2014) Characterization of human induced pluripotent stem cell-derived retinal pigment epithelium cell sheets aiming for clinical application. Stem Cell Reports. 2(2):205–218
25. McGill TJ, Stoddard J, Renner LM, Messaoudi I, Bharti K, Mitalipov S et al (2018) Allogeneic iPSC-derived RPE cell graft failure following transplantation into the subretinal space in non-human primates. Invest Ophthalmol Vis Sci 59(3):1374–1383
26. Chao JR, Lamba DA, Klesert TR, Torre A, Hoshino A, Taylor RJ et al (2017) Transplantation of human embryonic stem cell-derived retinal cells into the subretinal space of a non-human primate. Transl Vis Sci Technol 6(3):4
27. Little CW, Castillo B, DiLoreto DA, Cox C, Wyatt J, del Cerro C et al (1996) Transplantation of human fetal retinal pigment epithelium rescues photoreceptor cells from degeneration in the Royal College of Surgeons rat retina. Invest Ophthalmol Vis Sci 37(1):204–211
28. Ben M'Barek K, Habeler W, Plancheron A, Jarraya M, Regent F, Terray A et al (2017) Human ESC-derived retinal epithelial cell sheets potentiate rescue of photoreceptor cell loss in rats with retinal degeneration. Sci Transl Med 9(421):eaai7471
29. Hazim RA, Karumbayaram S, Jiang M, Dimashkie A, Lopes VS, Li D et al (2017) Differentiation of RPE cells from integration-free iPS cells and their cell biological characterization. Stem Cell Res Ther 8(1):217
30. Coffey PJ, Girman S, Wang SM, Hetherington L, Keegan DJ, Adamson P et al (2002) Long-term preservation of cortically dependent visual function in RCS rats by transplantation. Nat Neurosci 5(1):53–56
31. Lu B, Malcuit C, Wang S, Girman S, Francis P, Lemieux L et al (2009) Long-term safety and function of RPE from human embryonic stem cells in preclinical models of macular degeneration. Stem Cells 27(9):2126–2135
32. Iraha S, Tu HY, Yamasaki S, Kagawa T, Goto M, Takahashi R et al (2018) Establishment of immunodeficient retinal degeneration model mice and functional maturation of human ESC-derived retinal sheets after transplantation. Stem Cell Reports 10(3):1059–1074
33. Stanzel BV, Liu Z, Somboonthanakij S, Wongsawad W, Brinken R, Eter N et al (2014) Human RPE stem cells grown into polarized RPE monolayers on a polyester matrix are maintained after grafting into rabbit subretinal space. Stem Cell Reports 2(1):64–77
34. Plaza Reyes A, Petrus-Reurer S, Antonsson L, Stenfelt S, Bartuma H, Panula S et al (2016) Xeno-free and defined human embryonic stem cell-derived retinal pigment epithelial cells functionally integrate in a large-eyed preclinical model. Stem Cell Reports. 6(1):9–17
35. Lai CC, Gouras P, Doi K, Tsang SH, Goff SP, Ashton P (2000) Local immunosuppression prolongs survival of RPE xenografts labeled by retroviral gene transfer. Invest Ophthalmol Vis Sci 41(10):3134–3141
36. Del Priore LV, Ishida O, Johnson EW, Sheng Y, Jacoby DB, Geng L et al (2003) Triple immune suppression increases short-term survival of porcine fetal retinal pigment epithelium xenografts. Invest Ophthalmol Vis Sci 44(9):4044–4053
37. Mudumba S, Bezwada P, Takanaga H, Hosoi K, Tsuboi T, Ueda K et al (2012) Tolerability and pharmacokinetics of intravitreal sirolimus. J Ocul Pharmacol Ther 28(5):507–514

38. Koss MJ, Falabella P, Stefanini FR, Pfister M, Thomas BB, Kashani AH et al (2016) Subretinal implantation of a monolayer of human embryonic stem cell-derived retinal pigment epithelium: a feasibility and safety study in Yucatán minipigs. Graefes Arch Clin Exp Ophthalmol 254(8):1553–1565

39. Lin B, McLelland BT, Mathur A, Aramant RB, Seiler MJ (2018) Sheets of human retinal progenitor transplants improve vision in rats with severe retinal degeneration. Exp Eye Res 174:13–28

40. Hu Y, Liu L, Lu B, Zhu D, Ribeiro R, Diniz B et al (2012) A novel approach for subretinal implantation of ultrathin substrates containing stem cell-derived retinal pigment epithelium monolayer. Ophthalmic Res 48(4):186–191

41. Aramant RB, Seiler MJ (2002) Transplanted sheets of human retina and retinal pigment epithelium develop normally in nude rats. Exp Eye Res 75(2):115–125

42. Maeda T, Lee MJ, Palczewska G, Marsili S, Tesar PJ, Palczewski K et al (2013) Retinal pigmented epithelial cells obtained from human induced pluripotent stem cells possess functional visual cycle enzymes in vitro and in vivo. J Biol Chem 288(48):34484–34493

43. Westenskow PD, Kurihara T, Bravo S, Feitelberg D, Sedillo ZA, Aguilar E et al (2015) Performing subretinal injections in rodents to deliver retinal pigment epithelium cells in suspension. J Vis Exp 95:52247

44. Thieltges F, Liu Z, Brinken R, Braun N, Wongsawad W, Somboonthanakij S et al (2016) Localized RPE removal with a novel instrument aided by viscoelastics in rabbits. Transl Vis Sci Technol 5(3):11

45. Stanzel BV, Liu Z, Brinken R, Braun N, Holz FG, Eter N (2012) Subretinal delivery of ultrathin rigid-elastic cell carriers using a metallic shooter instrument and biodegradable hydrogel encapsulation. Invest Ophthalmol Vis Sci 53(1):490–500

46. Los LI, van Luyn MJ, Nieuwenhuis P (1999) Organization of the rabbit vitreous body: lamellae, Cloquet's channel and a novel structure, the 'alae canalis Cloqueti. Exp Eye Res 69(3):343–350

47. Banin E, Hemo Y, Jaouni T, Marks-Ohana D, Stika S, Zheleznykov S et al (2017) Phase I/IIa clinical trial of human embryonic stem cell (hESC)-derived retinal pigmented epithelium (RPE, OpRegen®) transplantation in advanced dry form age-related macular degeneration (AMD): interim results. Invest Ophthalmol Vis Sci 58(8):2320

48. Marmor MF (1990) Control of subretinal fluid: experimental and clinical studies. Eye 4(Pt 2):340–344

49. Maminishkis A, Amaral J, Charles ST, Bharti K, Miller SS (2016) Surgical tool for subretinal delivery of RPE implants. Invest Ophthalmol Vis Sci 57:12

50. Kamao H, Mandai M, Ohashi W, Hirami Y, Kurimoto Y, Kiryu J et al (2017) Evaluation of the surgical device and procedure for extracellular matrix–scaffold–supported human iPSC–derived retinal pigment epithelium cell sheet transplantation. Invest Ophthalmol Vis Sci 58(1):211–220

51. Wilson DJ, Neuringer M, Stoddard J, Renner LM, Bailey S, Lauer A et al (2017) Subretinal cell-based therapy: an analysis of surgical variables to increase cell survival. Retina 37(11):2162–2166

52. Seiler MJ, Aramant RB, Thomas BB, Peng Q, Sadda SR, Keirstead HS (2010) Visual restoration and transplant connectivity in degenerate rats implanted with retinal progenitor sheets. Eur J Neurosci 31(3):508–520

53. Mandai M, Fujii M, Hashiguchi T, Sunagawa GA, Ito SI, Sun J et al (2017) iPSC-derived retina transplants improve vision in rd1 end-stage retinal-degeneration mice. Stem Cell Reports. 8(1):69–83

54. Eberle D, Kurth T, Santos-Ferreira T, Wilson J, Corbeil D, Ader M (2012) Outer segment formation of transplanted photoreceptor precursor cells. PLoS One 7(9):e46305

55. Singh MS, Charbel Issa P, Butler R, Martin C, Lipinski DM, Sekaran S et al (2013) Reversal of end-stage retinal degeneration and restoration of visual function by photoreceptor transplantation. Proc Natl Acad Sci U S A 110(3):1101–1106

56. Silverman MS, Hughes SE, Valentino TL, Liu Y (1992) Photoreceptor transplantation: anatomic, electrophysiologic, and behavioral evidence for the functional reconstruction of retinas lacking photoreceptors. Exp Neurol 115(1):87–94
57. Aramant RB, Seiler MJ, Ball SL (1999) Successful cotransplantation of intact sheets of fetal retina with retinal pigment epithelium. Invest Ophthalmol Vis Sci 40(7):1557–1564
58. Woch G, Aramant RB, Seiler MJ, Sagdullaev BT, McCall MA (2001) Retinal transplants restore visually evoked responses in rats with photoreceptor degeneration. Invest Ophthalmol Vis Sci 42(7):1669–1676
59. Schwartz SD, Regillo CD, Lam BL, Eliott D, Rosenfeld PJ, Gregori NZ et al (2015) Human embryonic stem cell-derived retinal pigment epithelium in patients with age-related macular degeneration and Stargardt's macular dystrophy: follow-up of two open-label phase 1/2 studies. Lancet 385(9967):509–516
60. Mandai M, Watanabe A, Kurimoto Y, Hirami Y, Morinaga C, Daimon T et al (2017) Autologous induced stem-cell-derived retinal cells for macular degeneration. N Engl J Med 376(11):1038–1046
61. da Cruz L, Fynes K, Georgiadis O, Kerby J, Luo YH, Ahmado A et al (2018) Phase 1 clinical study of an embryonic stem cell-derived retinal pigment epithelium patch in age-related macular degeneration. Nat Biotechnol 36(4):328–337
62. Kashani AH, Lebkowski JS, Rahhal FM, Avery RL, Salehi-Had H, Dang W et al (2018) A bioengineered retinal pigment epithelial monolayer for advanced, dry age-related macular degeneration. Sci Transl Med 10(435):eaao4097

Chapter 7
3D Engineering of Ocular Tissues for Disease Modeling and Drug Testing

M. E. Boutin, C. Hampton, R. Quinn, M. Ferrer, and M. J. Song

Abstract The success rate from investigational new drug filing to drug approval has remained low for decades despite major scientific and technological advances, and a steady increase of funding and investment. The failure to demonstrate drug efficacy has been the major reason that drug development does not progress beyond phase II and III clinical trials. The combination of two-dimensional (2D) cellular in vitro and animal models has been the gold standard for basic science research and preclinical drug development studies. However, most findings from these systems fail to translate into human trials because these models only partly recapitulate human physiology and pathology. The lack of a dynamic three-dimensional microenvironment in 2D cellular models reduces the physiological relevance, and for these reasons, 3D and microfluidic model systems are now being developed as more native-like biological assay platforms. 3D cellular in vitro systems, microfluidics, self-organized organoids, and 3D biofabrication are the most promising technologies to mimic human physiology because they provide mechanical cues and a 3D microenvironment to the multicellular components. With the advent of human-induced pluripotent stem cell (iPSC) technology, the 3D dynamic in vitro systems further enable extensive access to human-like tissue models. As increasingly complex 3D cellular systems are produced, the use of current visualization technologies is limited due to the thickness and opaqueness of 3D tissues. Tissue-clearing techniques improve light penetration deep into tissues by matching refractive indices among the 3D components. 3D segmentation enables quantitative measurements based on 3D tissue images. Using these state-of-the-art technologies, high-

M. E. Boutin · M. Ferrer
National Center for Advancing Translational Sciences (NCATS), Rockville, MD, USA

C. Hampton · R. Quinn
National Eye Institute (NEI), Bethesda, MD, USA

M. J. Song (✉)
National Center for Advancing Translational Sciences (NCATS), Rockville, MD, USA

National Eye Institute (NEI), Bethesda, MD, USA
e-mail: minjae.song@nih.gov

© Springer Nature Switzerland AG 2019
K. Bharti (ed.), *Pluripotent Stem Cells in Eye Disease Therapy*,
Advances in Experimental Medicine and Biology 1186,
https://doi.org/10.1007/978-3-030-28471-8_7

throughput screening (HTS) of thousands of drug compounds in 3D tissue models is slowly becoming a reality. In order to screen thousands of compounds, machine learning will need to be applied to help maximize outcomes from the use of cheminformatics and phenotypic approaches to drug screening. In this chapter, we discuss the current 3D ocular models recapitulating physiology and pathology of the back of the eye and further discuss visualization and quantification techniques that can be implemented for drug screening in ocular diseases.

Keywords 3D in vitro model · Biofabrication · High-throughput screening · Human-induced pluripotent stem cell · Microfluidics · Optic cup · Organoid · Outer blood–retina barrier · Tissue clearing

7.1 Introduction

For years, the pharmaceutical and biotechnology industries have been incorporating major advances in scientific knowledge, "omics", engineering, and computational technologies to discover new disease targets, and develop the next generation of therapeutics. Despite the many scientific and technological advances of the last decades, the success rate from investigational new drug (IND) filing to drug approval has remained very low [1]. Over the past 60 years, the number of approved drug compounds per billion US dollar spent on research has continuously decreased, and the approval of a single drug now costs more than a billion dollars. Success rates have been reported for each phase of clinical trials: phase I (overall, 64%; ophthalmology, 84.8%), phase II (overall, 32%; ophthalmology, 44.6%), phase III (overall, 85.3%; ophthalmology, 77.5%), and Likelihood of Approval (overall, 10%; ophthalmology, 17%) [2, 3]. Most drug failures during phase II (48%) and III (55%) were due to a failure to demonstrate drug efficacy [4]. This high failure rate in drug development is in large part due to the use of overly simplistic in vitro cell assays and in vivo mouse models with limited predictive value during drug discovery.

Two-dimensional (2D) in vitro systems have provided assay models for studying healthy and diseased cellular and molecular mechanisms, and for the discovery of targets and biomarkers. These cellular systems have provided data on cytotoxicity of drugs [5], material for proteome/genome/transcriptome analysis [6], and a platform for gene editing [7]. However, results from these cellular systems must be used with caution because a 2D monoculture environment does not recapitulate the three-dimensional (3D) environment of in vivo tissue. Animal model systems, mostly murine and rabbit systems for ocular pathologies [8] and development [9], have been critical for the understanding of disease biology and drug testing in a more relevant in vivo setting. With the advent of biologics in the pharmaceutical field, these animals have helped establish the efficacy of several anti-vascular endothelial growth factor (VEGF) antibodies for the treatment of wet-form age-related macular degeneration (AMD) as well as other neovascularizing pathologies [10]. The combination of 2D in vitro and in vivo models has been the gold standard for drug development in the preclinical stages. However, due to the many physiological differences between model animals and humans, the results of animal studies often fail to translate to human trials. Humanized animal models have been developed to minimize the discrepancies between species by transfecting human genes or introducing human hematopoietic stem cells (HSCs) into immunodeficient mice, which has produced promising models of human infectious disease [11]. In parallel, the development of co-culture systems containing multiple cell types, in many cases generated using primary or induced pluripotent stem cell (iPSC)-derived cells, has initiated a paradigm shift for in vitro systems, in which researchers have also moved from 2D to 3D assays. Additionally, the field of microfluidics has led to the generation of new in vitro systems by providing mechanochemical cues to 3D cellular components. Hence, two major fields, microfluidics and tissue engineering, are becoming mainstays in the development of 3D in vitro systems by providing dynamic and 3D microenvironments.

Microfluidic "chips" contain micro/mesoscale (100–500 μm width) fluidic channels which provide mechanical cues and a dynamic milieu to cells. One of the first applications of in vitro microfluidics systems was to mimic the vascular system. Vascular endothelial cells (ECs) are exposed to hemodynamic forces in the human physiological system (venous system, 1–6 dyn/cm^2; arterial system, 10–70 dyn/cm^2) [12]. It has been shown that flow shear stress is critical to the structure and function of ECs, according to the study of an EC monolayer within a fluidic channel/chamber. This study showed shear flow-induced cytoskeletal rearrangements and significant mechanotransduction by activating transmembrane proteins (e.g., VEGF receptor, ion channel, integrin, PECAM-1) [13]. The addition of a fluidic channel alongside this "vascular" channel, with a porous membrane separating these two channels, enabled a dynamic co-culture system, so-called "organ-on-a-chip", to study interfaces between two mechanobiological regimes [14, 15]. The addition of hydrogel material in the microfluidic system further advanced the structure of the "vascular" channel by providing a 3D microenvironment where multiple cell types can be introduced. Fibroblasts and pericytes, major stromal cells, have been previously reported to play a crucial role in EC's viability. Fibroblasts secrete not only extracellular matrix (ECM) proteins but also VEGF, which is essential for vessel viability and development [16]. Pericytes, perivascular cells, secrete angiopoietin-1 (ANG-1), which is a ligand of the EC TIE-2 receptor. ANG-1 is a well-known factor involved in the regulation of angiogenesis in human development and pathology [17, 18]. Inclusion of these cells within the hydrogel space promotes EC tube formation and provides the 3D architecture of "vascularized" tissue. The self-assembled tubes of ECs within a fibrin-based hydrogel-filled chamber (~1 mm^2) are connected to the adjacent microfluidic channels where an EC monolayer was performed on the wall. The 3D "vascular" network in the chamber has been successfully perfused through anastomoses between ECs from the chamber and ECs from the adjacent channel. Perfusion and connectivity were confirmed by introducing dextran into the adjacent channels [19–23].

Tissue engineering technologies that bridge the gap between 2D in vitro and in vivo systems have grown rapidly in recent years. Bioprinting is a tissue engineering tool that provides a 3D microenvironment for cell/tissue growth by generally dispensing a mixture of hydrogel with cells, called "bioink". Rapid, scalable, and reliable fabrication of architecturally and physiologically defined functional human tissues still remains a challenge which bioprinting promises to address by integrating five recently emerging technologies: (1) 3D bioprinters are now available with precise XYZ control to reproducibly fabricate tissues with defined 3D geometries, (2) biomaterial and biocompatible hydrogels are being developed to support the 3D structure of tissue-embedded cells, (3) bioreactors are now being designed to deliver mechanical and chemical cues to the cells seeded in the printed tissue, (4) it is also now possible to obtain autologous cells from patients, including human iPSCs, and finally, (5) the ability to quantitatively characterize the morphology and functionality of printed tissues is also now possible using noninvasive technologies such as high-resolution fluorescence confocal/multiphoton microscopy and image analysis techniques.

In ophthalmology, while there are multiple 3D systems currently available, few studies have demonstrated applications of the systems which mimic human ocular physiology. In this chapter, we discuss current options to 3D model ocular tissues using the aforementioned technologies and to authenticate engineered ocular tissues. We further discuss the current challenges in modeling diseases and drug screening of ocular diseases.

7.2 3D Modeling of the Back of the Eye: Outer Blood–Retina Barrier

7.2.1 Background

The back of the eye is a light-sensing layered tissue composed of neurosensory retina, retinal pigment epithelium (RPE), and choroid. Given its unique structure and location between the choroid and retina, the RPE acts as a major homeostatic unit, regulating both choroid and retina tissues. For example, the RPE phagocytoses shed photoreceptor rod outer segments to protect retina, forms tight junctions to prevent large molecules from infiltrating into retina from the blood, and secretes growth factors in a polarized manner to maintain choriocapillaris. Pigment epithelial derived factor (PEDF), an antiangiogenic factor, is predominantly secreted in the apical region, whereas VEGF, a proangiogenic factor, is predominantly secreted in the basolateral region [24–26]. Imbalance of the secretion pattern is thought to initiate choroidal neovascularization (CNV), which is the pathological condition of excessive growth of choriocapillaris toward RPE. RPE cell death debilitates its phagocytic capability, which maintains the viability of photoreceptors. RPE cell death also breaks the mechanical integrity of the RPE barrier by disrupting intercellular junctional proteins such as ZO-1, occludin, cadherins, and β-catenin [27]. The disruption of intercellular junctions disturbs apicobasal polarity, which regulates the microenvironment of both apical and basal sides by controlling the secretion of distinct molecules at each side. It has been previously reported that, once tight junction and cell polarity are disrupted, the protein secretion pattern of the RPE and the microenvironment of choriocapillaris and photoreceptors are significantly changed [26].

7.2.2   3D Modeling of the Back of the Eye: Engineered Systems

Hamilton et al. showed RPE–EC interactions in an outer BRB model by forming a monolayer of ARPE-19 on one side of an amniotic membrane and forming an HUVEC monolayer on the other side of the membrane in a transwell format [28]. The model demonstrated barrier function, as measured by transepithelial electrical resistance (TEER), and EC fenestration, which is a key feature of choriocapillaris, by transmission electrical microscopy (TEM). Increasing the complexity of the choroid tissue, Chung et al. developed a model containing 3D vascularized "choroid" tissue within a microfluidic channel adjacent to two additional channels [23]. Acellular fibrin-based hydrogel was introduced in the middle channel providing spaces for angiogenesis from the "vascular" channel and facilitating quantification of angiogenesis toward RPE cells (ARPE-19) of the third channel. By increasing VEGF concentration (10–100 ng/ml) in the RPE channel, directional angiogenesis toward the RPE channel was demonstrated. Although the method of creating the VEGF gradient was not physiologically relevant, this was the first model mimicking CNV of wet AMD. The model was further tested with therapeutic intervention using an anti-VEGF drug, Bevacizumab (0.3 mg/ml), which is the gold standard for treatment of CNV [23]. Permeability was assessed by GFP and RFP dextrans to EC and RPE channels, respectively.

We recently developed an outer BRB model using bioprinting technology with iPSC-derived ECs and RPEs (Fig. 7.1). The iPSC technology provides extensive access to human cells as it enables the reprogramming of human somatic cells into pluripotent stem cells. Bioprinting of the fibrin-based gel containing fibroblasts, pericytes, and iPSC-derived ECs, called "bioink", formed choroid-like 3D vascularized tissue on an electrospun biodegradable polymer scaffold. The scaffold provided mechanical support to the bioprinted tissue until stromal cells secreted sufficient amounts of ECM proteins. Human iPSC-derived RPE cells formed a monolayer on the other side of the scaffold. This model demonstrated fenestrated capillaries, and barrier function was assessed by measuring TEER [29, 30, 31].

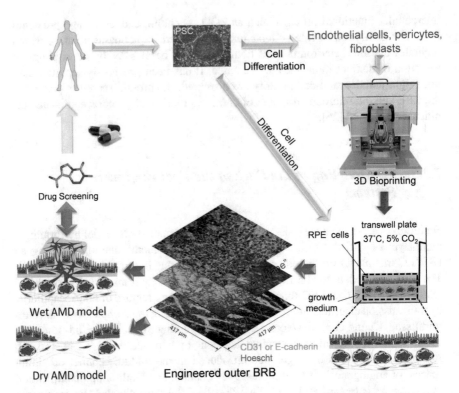

Fig. 7.1 3D Bioprinting of outer BRB using iPSC derived cells: modeling age related macular degeneration for drug discovery

7.2.3 3D System Validation

During the visual cycle, 11-cis-retinal is photoisomerized to all-trans-retinol in photoreceptors. In RPE cells, all-trans-retinol is reversed to 11-cis-retinal by retinoid isomerohydrolase (RPE 65), and the converted 11-cis-retinal is transferred to the photoreceptor for phototransduction. RPE 65 is widely used for identifying RPE, and this can validate RPE location within the engineered BRB tissue. Polarized morphology of RPE can be confirmed by immunostaining for Ezrin, which is a linker protein between the plasma membrane and cytoskeleton. Ezrin expression is localized on the apical region of RPE when it is fully polarized. RPE tight junctions are the major contributor to RPE barrier function, and tight junctions of the RPE can be visualized by ZO-1, Occludin, and Claudin19. TEM visualizes apical processes, polarization, pigmentation, basal infolding, and tight junctions. Barrier function of the RPE monolayer can be assessed by measuring TEER or permeability of dextran [23, 32, 33].

Bruch's membrane is the basement membrane located between RPE and choroid. Bruch's membrane is composed of multiple layers of ECM: RPE basal lamina, inner collagenous layer, elastic layer, outer collagenous layer, and choriocapillaris basal lamina [34]. Since it consists of multiple types of ECM, the presence of the essential ECM components for in vitro models need to be confirmed.

Choriocapillaris is a key component in the choroid for gas exchange and nutrient delivery. The fenestrated features of choriocapillaris leads to extensive gas exchange between RPE and blood. The fenestration can be tested by expression of the fenestration marker, fenestrated endothelial cell linked protein (FELS). However, the expression of FELS does not provide sufficient evidence of fenestrated morphology. Visualization of ultrastructure by electron microscope is suggested for the final confirmation [35]. EC biomarkers such as PECAM-1, VE-cadherin, and Isolectin-B4 can be used to visualize 3D tube formation. 3D reconstruction of confocal images or histological cross sections can confirm the hollow structure of the EC tubes.

7.3 3D Modeling of the Back of the Eye: Pluripotent Stem Cell-Derived 3D Retina Organoids

7.3.1 Background

Both classical experiments and recent progress in embryology have offered insight into the genetic underpinnings of the stepwise development of the eye and neurosensory retina. Greater understanding within this body of knowledge in conjunction with advances in stem cell technology (specifically the generation of iPSCs [36]) has provided opportunities to engineer signaling cascades for the differentiation of pluripotent cells. In this way, it becomes possible to model various developmental stages of the neurosensory retina.

Developmentally, the eyes are bilateral extensions of diencephalon that begin as optic cups on day 23 of gestation. Between day 25–35, these optic cups evaginate and extend toward overlying surface ectoderm through mesenchyme [37]. Inductive signaling occurs here between the optic vesicle and the surface ectoderm, which eventually becomes the lens placode. More superficial surface ectoderm is fated to become cornea. This evaginating optic vesicle has dorsal-ventral pattern specification, with RPE developing dorsally, optic stalk developing ventrally, and neurosensory retina between the two. Subsequent invagination of the optic vesicle and lens placode leads to formation of the bilayered optic cup, which begins to resemble the ultimate shape of the eye. At this time point, the regions of presumptive neural retina and RPE that were bipotential start to become fated. RPE and neural retina patterning occurs over week 5, with neurogenesis thereafter. Progenitors rapidly proliferate with subsequent differentiation, migration, and organization from weeks 7 to 24. Ganglion cells, cones, and horizontal cells are first to appear first while rods, bipolar cells, and muller glia come later. Development of one cell type leads to production of the subsequent types in sequence secondary to local signaling mechanisms. Maturation and outer segment formation appears from weeks 18 to 28.

This process is mediated by a self-regulatory transcriptional network. Eye-field transcription factors include PAX6, RX, LHX2, SIX3, and SIX6 [38], with expression of OTX2 preceding neural induction. Activin-like factor induces microphthalmia-associated transcription factor (MITF) and RPE specification, while activation of the ERK pathway via fibroblast growth factor (FGF) from the lense ectoderm induces VSX2 (CHX10) and SOX2, promoting retinal development. Simultaneously, transcriptional repression limits the potentiality as development proceeds.

7.3.2 Pluripotent Stem Cell-Derived Optic Cup

With the use of factors to manipulate the signaling cascades associated with these networks, it becomes possible to manipulate the fates of stem cells in culture. There has been significant progress in both 2D and 3D culture methods for obtaining cells fated to a retinal lineage. For example, high efficiencies (80%) of retinal progenitor cells (primarily ganglion and amacrine cells) can be generated from human embryonic stem cells (ESCs) by first antagonizing the bone morphogenic protein (BMP) and Wnt pathways while treating with IGF-1, then treating with FGF [39]. This method resulted in expression of CRX, the earliest known photoreceptor marker, in 10 days of culture. The genetic profile of cells cultured under these conditions showed an accelerated rate of maturation, with 25 days of culture resembling that of the human fetal day 91 retina.

Inhibition of Wnt and Nodal before treatment with retinoic acid and taurine in a stepwise fashion allows increasing maturity, specifically identification of rods and cones [40]. With this culture method, by day 35, around 15% of colonies express RX and PAX6. CRX expression was observable within progenitor colonies by day 90; however, extension of the culture duration to 120 and 170 days led to increasing expression of CRX and mature markers like rhodopsin, recoverin, red/green opsin, and blue opsin. Manipulation of the culture conditions, and exposing cells to antagonists and cytokines that direct pluripotent cells to a ventral, eye field, and photoreceptor specific fates, results in higher differentiation efficiencies [41]. Further modification, with the use of RGD hydrogels can increase optic vesicle formation within embryoid bodies [42].

Even greater maturation of retinas-in-a-dish was demonstrated with the self-organizing optic cups from both mouse and human ESCs [43, 44]. With a serum-free culture of embryoid body-like aggregates with quick aggregation and a basement membrane component, which is either matrigel or laminin and entactin [later replaced with a transient BMP4 treatment, [45]], aggregates of stem cells can become stratified retinal tissue in a dish. Slight modifications to the protocol developed for mouse ESC (high knockout serum replacement enhancing retinal epithelium formation, Wnt inhibition to prevent caudalization, and addition of fetal bovine serum and sonic hedgehog to augment retinal differentiation) allow for human ESCs to form homogenous stem cell aggregates that first form hollow spheres.

These latter develop hemispherical vesicles evaginating from the main body that develop into a two-walled optic cup-like structures. Treatment with Wnt agonists to induce RPE differentiation during day 15–18 enhances MITF expression without suppressing CHX10/VSX2 and allows for RPE/neural retina interaction. The distal portion of the vesicle then invaginates from day 19 to 24 expressing VSX2 (CHX10), SIX3, RX, and PAX6, while proximal cells express PAX6, MITV, and OTX2. Treatment with Activin leads to significant increase in RX expression.

Furthermore, these tissues exhibited interkinetic nuclear migration with progenitors maturing into photoreceptors, ganglion cells, bipolar cells, horizontal cells, amacrine cells, and muller glia. Similar to embryogenesis, ganglion cells appeared first around day 24 at the basal zone of the neural retina. CRX-positive photoreceptor progenitors were present in these earlier time points, but relatively sparse. However, by day 60, the engineered retina contained recoverin-expressing photoreceptors that become substantially denser after day 91. Additionally, the rod-specific marker neural retina leucine zipper (NRL) and the early cone marker retinoid X receptor (RXR) gamma appeared after day 126. Further modification of the protocol with removal of antibiotics and feeder cells was later demonstrated to increase maturation, as observed through the formation of outer segment-like structures [46]. Trisecting of aggregates was shown to enhance yield [47], which would be beneficial for potential transplantation procedures as photoreceptor integration has been demonstrated in mouse models [48].

However, even in the absence of exogenous factors, human anterior neuroectodermal ESCs mature into a population expressing eye-field transcription factors. These can subsequently grow into neural rosettes on a laminin-coated culture dish before being lifted and grown as neurospheres, which acquire a retinal fate and developmentally mimic normal retinogenesis [49, 50]. In this stepwise differentiation, pluripotency is rapidly lost. By day 2 of differentiation cells expressed both PAX6 and RX, and by day 10 LHX2, OTX2, and SOX1. Maintenance of the cultures showed increasing MITF expression from day 16 to 37 and coexpression of VSX2 (CHX10) by day 40. Similar to embryonic retinogenesis, these bipotential populations of cells became subsequently fated by day 50, expressing exclusively either MITF or VSX2 (CHX10). CRX, recoverin, and cone-specific opsin were identified within neurospheres by day 80. Electrophysiological testing of recoverin-positive cells showed a current–voltage relationship consistent with mammalian photoreceptors. This culture method supports the endogenous signaling mechanisms that allow for the recapitulation of retinogenesis and maturation of the retina.

Exploitation of these intrinsic cues occurring between stem cells can allow for the generation of mature retinal structures such as the outer segment and behaviors such as responsiveness to light [51]. Culture of pluripotent stem cells in a neural induction medium first leads to development of anterior neuroepithelial cells expressing PAX6 and SOX2 with subsequent LHX2 from retinal progenitors by day 8. Concomitant expression of VSX2 (CHX10) and MITF illustrates the bipotential nature of the progenitors; however, similar to development, these cells become restricted. These regions expressing PAX6, LHX2, RX, and VSX2 (CHX10) segregate to neural retina and those expressing MITF and PAX6 to RPE. The tissue then begins to resemble the developmental optic cup from day 17 to 25 with subsequent center-to-periphery neurogenesis occurring over days 35–49 with maturation of photoreceptors, amacrine cells, and horizontal cells expressing OTX2, AP2α, and PROX1. Maintenance of the culture until day 147 allows for cells to segregate into a well-organized outer nuclear layer, expressing more mature markers like rod opsin, L/M-opsin, and S-opsin. Outer segments begin to appear around day 175, with cells expressing proteins involved in phototransduction including the α-subunit of rod transducin (GT1α), the α- and β-subunits of the rod cGMP-phosphodiesterase (PDE6αβ), the rod cyclic-nucleotide-gated channel α-subunit and β-subunit, and retinal guanylate cyclase 1. One limitation inherent in using intrinsic cues for retinal differentiation and maturation is the variability in the differentiation capacity of various cell lines. However, the addition of Wnt antagonists enhances the ability of progenitors to self-organize [52] and can allow for more efficient generation of retinal tissue.

Alternative protocols use sheets of human ESCs suspended in matrigel and cultured in suspensions. First, cells are differentiated toward a neural fate with N2B27 medium, then, in the absence of extrinsic patterning cues, mature to retinal progenitor cells that express eye-field transcription factors by day 10 (SIX3, PAX6, RAX, OTX2) and markers of photoreceptor progenitors by day 13 [53, 54]. Cells cultured in this way first form epithelialized cysts, which spontaneously attach to the culture surface. These can be isolated with dispase and grown in suspension during days 12–17 to enrich VSX2 expression. Long-term culture of these organoids results in stratified neural retina with mature photoreceptor markers like rhodopsin and recoverin by day 152, and outer segments by day 187.

There have been a variety of published protocols regarding the differentiation of mature retinal tissue. Key differences between methods include the application of exogenous factors ultimately leading to differences in the ease of the protocol, and the time at which cell types are acquired. All result in the ability to generate mature human retinal tissue for disease modeling, opening the door for new studies in teasing out pathological mechanisms of disease, developing platforms for drug discovery, and ultimately tissue transplantation. In fact, exploratory transplantation procedures in mouse and monkey models have shown some success, with possible host-graft synaptic connections [55] and some recovery of light-responsive behaviors in diseased animals [56].

7.4 Visualization and Quantification of Markers in 3D Tissue Models

7.4.1 Traditional Histology

Historically, assessment of tissue morphology has been performed by collecting thin physical sections of a tissue and staining to improve visualization of the structures of interest. Tissues are sectioned by embedding in a wax paraffin block for paraffin sectioning or optimum cutting temperature medium for cryosectioning. For visualization of the retinal layers, the most commonly used stain is the colorimetric stain hematoxylin and eosin (H&E), which colors cell nuclei blue and the protein-rich cytoplasm and extracellular matrix shades of pink. Additionally, specific proteins can be labeled through fluorescent immunocytochemical protocols or colorimetric immunohistochemical protocols.

A consideration when performing staining in retinal tissue is the presence of melanin pigmentation within the RPE and choroid layers. This pigment is similarly colored to the commonly used immunohistochemical chromogen DAB (3,3′-Diaminobenzidine), making it difficult to distinguish DAB labeling within pigmented tissues. For immunofluorescently stained samples, pigment may prevent signal collection from pigmented areas by absorbing light. Notably, melanin pigmentation can also interfere with genomic analysis, such as RNA/DNA quantification and the polymerase chain reaction [57, 58]. Due to these challenges, methods have been developed to decrease pigmentation in tissues through melanin bleaching, using chemicals such as potassium permanganate/oxalic acid or dilute hydrogen peroxide. While most of these protocols have been developed for melanin present in skin or melanoma tissue [59, 60], they are also effective in retinal tissue [61, 62]. Melanin bleaching protocols are performed following tissue sectioning, prior to immunostaining protocols.

7.4.2   Tissue Clearing

While traditional histology is effective for visualizing layered tissue morphology such as the retina, it is difficult to understand the 3D organization of a tissue from thin sections. Fluorescent confocal and two-photon imaging provide alternative solutions to acquire 3D images from intact tissues. However, these imaging techniques encounter technical issues due to the opacity of dense tissues. Light scatters as it travels through cellular compartments with different refractive indices and through extracellular matrix. This scattering makes the collection of fluorescent signal during imaging challenging, especially in 3D tissues. While imaging ability varies widely between different tissue types, this scattering phenomenon typically leads to an inability to image deeper than several cell layers into tissues (~50–100 μm). Additionally, in retina and skin, melanin pigmentation absorbs light and further prevents tissue imaging.

Different techniques have been developed to decrease light scattering within 3D samples and increase the amount of fluorescent signal recovered. Optical tissue clearing is a technique that aims to homogenize refractive index mismatches within tissues and decrease scattering caused by extracellular matrix components. During the past 5–10 years, the number of published protocols has dramatically increased, spurred by improvements in microscope objectives and analysis capabilities (for comprehensive reviews, please see [63] and [64]).

Clearing techniques can be broadly grouped into reversible, chemical protocols and irreversible, crosslinking protocols. Within tissues, a large difference in refractive indices exists between proteins and lipid bilayers (RI ~ 1.50, [65]), and cellular compartment mediums (RI ~ water 1.33). The reversible chemical clearing protocols aim to decrease the refractive index differential within tissues by incubating in reagents with high refractive indices. Some protocols use high refractive index solutions such as organic solvents (BABB [66]; 3DISCO [67]), or concentrated sugar solutions (SeeDB [68]). Other protocols incubate in high refractive index aqueous solutions and additionally reduce the scattering of lipid and protein components with reagents such as urea and triton X-100 (Scale [69]; ScaleS [70]; CUBIC [71]). The first published crosslinking clearing technique, termed CLARITY, achieves its effect by crosslinking tissue proteins to an acrylamide gel and using detergent to remove tissue lipids, either passively over an extended time scale, or actively by pulling the detergents using electrophoresis [72]. As lipids are one of the main light scattering components of tissues, this method has been successful in enabling imaging deep into different tissue types, and many new protocols have been developed based on this approach [73–76].

The choice of clearing reagent and protocol must be designed carefully, considering characteristics such as the tissue type of interest, the refractive index of the tissue type, sample size, necessary imaging depth, macromolecules of interest, reagent cost, time/throughput requirements, and fluorophore compatibility. For example, for small (<500 μm thickness) tissue samples, the complex and involved steps of the CLARITY acrylamide crosslinking protocols might not be necessary for imaging. To visualize an endogenous fluorophore or fluorescent antibody, some of the organic solvent protocols should be avoided, as they have been shown to quench fluorophores. If the macromolecule of interest is a lipid, the delipidation CLARITY protocols and perhaps the Scale and CUBIC protocols would be avoided, whereas if the macromolecule of interest was a protein, compatibility would need to be checked with the urea-containing Scale protocol. Most protocols take on the order of days-weeks, depending on the protocol and the sample being cleared. Higher-throughput protocols have been developed to clear samples in a shorter time frame with fewer steps [77–80].

As described above, due to the light-absorbing properties of pigmentation found in the RPE and choroid layers, retinal tissue adds an additional hurdle to fluorescent imaging of 3D tissue samples. The aforementioned melanin bleaching strategies for sectioned tissues can also be adapted to 3D retinal tissue and enhance deep tissue imaging [61]. Bleaching protocols are very effective at removing the dark melanin pigmentation; however, in 3D samples, the opaque tissue of the eye remains. To achieve a sample with minimal light scattering, we have found that the combination of bleaching, followed by immunocytochemical staining and optical clearing, results in optically transparent and imageable tissue.

7.4.3 3D Segmentation and Quantification

Improvements in clearing and microscopy technology have led to an increased ability to resolve fine details within 3D samples, and a push to develop more advanced analysis methods for 3D tissues. For screening, the most common measurements currently taken of 3D tissues are tissue size by brightfield analysis, and overall fluorescence by widefield microscopy, assessments which do not begin to analyze the wealth of 3D information within the tissue structure. Initial work to improve analysis strategies in 3D have begun by improving segmentation of cell nuclei, a task complicated by light scattering and by the differences in resolution between the x, y, and z imaging axes. Several protocols have demonstrated highly accurate segmentation of cell nuclei in 3D samples [81–83]. However, these protocols have not yet been applied to large data sets. In an effort to develop nuclear segmentation protocols that efficiently analyze large data sets, our group recently developed a protocol that combines optical clearing of 3D tissues, high-content imaging, and high-throughput nuclear segmentation analysis [78]. Using this analysis, we performed nuclear segmentation in an automated fashion for a 384-well plate of spheroids in 5–6 h. By creating the analysis such that it can be run in parallel, we also improved the processing speed by analyzing in parallel on hundreds of cloud-based cpu workers. Protocols such as these can be used to analyze nuclear fluorescent signals, for example, markers of DNA damage, promoter activity, and transcription factors and can be built upon to analyze cytosolic labeling.

Machine learning is a type of artificial intelligence that is being rapidly applied within the field of high-content image analysis. These analyses typically use a sample image data set to train, identify patterns, and learn about the data behavior. Based on this control data set, the program then makes predictions on novel data sets. Machine learning methods have been used successfully in 2D data sets, for example, to classify and score cellular subtypes [84]. Ultimately, the combination of 3D segmentation and analysis protocols with innovative machine learning technologies will enable sophisticated analyses that can adapt to different 3D data sets.

7.5   Drug Screening of Ocular Tissue Models

7.5.1   High-Throughput Drug Screening (HTS) with Stem Cell-Derived Ocular Disease Model

The development of 3D ocular tissue models using iPSC-derived cells and clearing/imaging techniques described above provides a unique opportunity to implement drug screening for ocular diseases using assay systems that are more physiologically-relevant. It is expected that using patient iPSC-derived cells in the context of a relevant tissue, where native physiological cell–cell and cell–microenvironment interactions are maintained, will yield drug responses that are more similar to those obtained in humans, and as a consequence, these assays systems will be more predictive of drug responses in clinic. In the field of drug screening, assay robustness, reproducibility, and sample throughput are key aspects that determine the stage in the drug discovery pipeline at which an assay can be used [85]: primary screening of chemically diverse collections for lead identification for new drug discovery programs (100,000–1 M compounds) and focused collections for drug repurposing or annotated collections for chemical biology studies (1000s compounds); secondary assays for HTS hit validation or optimization during medicinal chemistry efforts (100–1000s compounds); and tertiary assays for lead optimization and validation (10–20s compounds). Assays used for primary screens are in most cases developed in 384- or 1536-well microplate format in order to increase throughput by using laboratory automation, and reduce cost of reagents because of the smaller volumes used. Because large number of cells are needed to implement primary screens (at least 1000–5000 cells per well in 1536-, and 5000–20,000 cells per well in 384-well), cell production, scale-up, homogeneity of the cell population, and stability of responses with passage and time are critical to have robust and scalable cell-based assays. For this reason, cell lines engineered to express reporters such as GFP or luciferase are widely used in primary high-throughput screens. There are many examples of drug screens using iPSC-derived cells, both to increase the efficiency of differentiation protocols as well as for the correction of disease phenotypes [86]. In many cases, these screens use HTS friendly readouts such as cell viability and reporters, which are engineered at the ESC (mostly from transgenic mice) or iPSC stage of the cells.

High content screening (HCS) is the leading methodology to implement this type of phenotypic screen [87, 88], in which the levels of relevant physiological markers of interest are quantitated either by fusing a protein to a fluorescence reporter, or by immunostaining, followed by cell imaging using automated fluorescent microscopes, coupled with data analysis using algorithms designed to quantify one or more specific phenotypes of interest [89]. One example of a reporter screen has been described for the identification of compounds that accelerate photoreceptor differentiation by stimulating expression of CRX and, consequently, its downstream targets [90]. The strategy applied for this HTS was to use a CRXp-GFP H9 human ESC line. This group showed that human ESC can differentiate and self-organize in vitro to 3D retina cultures in a sequential process starting with floating aggregates of human ESC, which form optic vesicles where neural retina and pigment mass develop. Transcriptomics analysis during the process of differentiation and optical cup formation allowed for the identification of specific sets of genes expressed at the different stages of differentiation. With that information, fluorescent reporters controlled by promoters of such genes can help identify defined developmental stage(s) or individual cell types. Expression of CRXp-GFP reporter in differentiating human ESCs provided the reading output for determining whether a compound accelerated photoreceptor differentiation. The group adapted the assay to 1536-well microplate HTS and went on to screen three-focused collections, the library of pharmacologically active compounds (LOPAC, 1280 compounds), a NIH mechanism interrogation plate (MIPE) collection of annotated compounds (1900), and a drug repurposing collection (~ 3000 compounds). The authors indicated the development of a panel of reporter cells at different stages of photoreceptor differentiation for future drug screening. In another example, Fuller et al. [91] described the use of a rhodopsin (Rho)-GFP-engineered primary retinal neuron from transgenic mice to identify molecules that promoted photoreceptor differentiation. In this example, a hRhoGFP knock-in mouse was generated by replacing the mouse rho open reading frame with a human Rho-GFP fusion construct, thus creating a rhodopsin reporter controlled by native regulatory mechanisms. The primary retinal neurons extracted from the mice were seeded in a 1536-well plate, and after 21 days, photoreceptors in the culture developed bright GFP fluorescence with bright punctate labeled as "peak rhodopsin" objects, believed to be proto-outer segments. This assay was used to screen the LOPAC collection. These are two examples of focused drug screening using ESC- or iPSC-derived cells to identify compounds that enhance differentiation for potential therapeutic use, both as a treatment to induce photoreceptor regeneration in vivo and also as reagents to enhance the production of iPSC-derived cells for cell-based therapies. In the work by Fuller et al. [91], they also describe how the same cell system can be used to screen for compounds that correct a disease phenotype.

A cell viability assay after stress-induced photoreceptor cell death was developed to identify compounds that were neuroprotective by preventing the slow but progressive loss of photoreceptors and thus could be developed as potential treatments for diseases of retinal degeneration. An HTS friendly readout such as CellTiterGlo, which measures cellular ATP as a surrogate of cell viability, was used to screen the LOPAC collection. In a similar approach, Chang et al. [92] reported a cell viability assay using AMD-patient iPSC-derived RPE cells in which damaging oxidative stress was induced with H_2O_2 treatment. Chronic oxidative stress induces RPE damage that is responsible for the aging process and plays a significant role in the pathogenesis of AMD. These patient-derived RPEs with the AMD-associated background (AMD-RPEs) exhibited reduced antioxidant activity compared with normal RPE cells. As a proof of concept, the authors tested a small set of dietary supplements for retinal protection and natural compounds for antioxidant effects using a methyl thiazol tetrazolium (MTT) assay to measure cell viability. Among several screened candidate drugs, curcumin caused the most significant reduction of reactive oxygen species (ROS) in AMD-RPEs and protected these AMD-RPEs from H_2O_2-induced cell death.

7.5.2   3D Organoids: Human iPSC-Derived Retinal Cup

The examples above illustrate the development of ESC or iPSC-derived assay systems that are amenable to drug screening. While there are not yet studies screening collections of compounds using the recently developed 3D retinal cup organoids, there have been reports of work towards the validation of these systems for screening, including developing HTS friendly assays, extensive morphological, physiological, and pharmacological characterization and validation, and streamlining of iPSC production protocols [52]. Several proof-of-concept studies have been published which demonstrate how 3D retinal cup models can be used as systems to discover new therapeutic interventions. In an approach similar to that described above, induced photoreceptor cell death in mouse iPSC-derived 3D retinal organoids (3D-retinas) by 4-hydroxytamoxifen (4-OHT), which induces photoreceptor degeneration in mouse retinal explants, was established, and then a live-cell imaging system to measure degeneration-related properties was developed to quantify the protective effects of representative ophthalmic supplements for treating the photoreceptor degeneration [93]. In this work, GFP expressing optic cups were prepared from mouse Nrl-GFP iPSCs. Real-time measurements of total intensity from the GFP expressing optic cup using an InCuyte Zoom (Essen BioScience, Inc) enabled the quantitation of 4-OHT-induced acute cell death in the ERRβ-expressing rod photoreceptors. In order to validate the pharmacological relevance of this assay, vitamin E, which is a supplement known to prevent 4-OHT-induced photoreceptor cell death, and lutein, astaxanthin, and anthocyanidin, known to have protective effects against light-induced photoreceptor degeneration in mouse native retina, which is considered to model AMD and Retinitis Pigmentosa (RP), were tested. The results confirm that while vitamin E was able to prevent 4-OHT-induced photoreceptor cell death, the other compounds did not, as was expected from their protective effect being for a different photoreceptor degeneration mechanism. In a similar approach, GFP and YFP reporters were engineered in hiPSC cells, and fluorescence reads were used to measure the effects of H_2O_2 ROS induction on the viability of 3D retina organoids, while a mitochondrial membrane potential fluorescent probe was used to quantify the effects of carbonyl cyanide m-chlorophenyl hydrazone (CCCP), a mitochondrial uncoupler [94]. This 3D automated reporter quantification (3D-ARQ) assay platform was extensively validated for HTS robustness in a 96-well plate format. In another published work [95], 3D human iPSC-derived optic cups were used as disease models of Leber Congenital Amaurosis (LCA), an inherited retinal dystrophy that causes childhood blindness. In LCA, photoreceptors are especially sensitive to an intronic mutation in the cilia-related gene CEP290, which causes mis-splicing and premature termination. LCA optic cups were created using iPSCs with the common CEP290 mutation. Even though these iPSCs differentiated normally to 3D optic cups and RPE, the highest levels of aberrant splicing and cilia defects were observed in optic cups. In this case, they did not use compounds to validate the pharmacological value of this model but showed that treating the optic cups with an antisense morpholino oligomer effectively blocked aberrant splicing and restored expression of full-length CEP290, restoring normal cilia-based protein trafficking. In this case, the readouts used to quantitate the correctional effect of the morpholino on the disease phenotype were immunostaining and RT-PCR-based, which are complex to implement for drug screening, and therefore, even though the system was validated as disease model that could be corrected with a treatment, efforts will still need to be made to further develop the system for drug screening.

7.5.3 HTS Readouts and Validation

The complex nature of 3D tissue systems, including scale-up of production and complexity of the disease-relevant readouts, limits their adaptation to HTS and sample throughput. These systems are currently limited to hit validation [96], pharmacological validation of the models, and a few focused screens. Most of the reported

3D tissue model screens have used tumor spheroids [97] or patient-derived tumor cells self-organized organoids [98], and have screened small, focused collections of compounds as proof-of-concept screens, in most cases using cell viability assays or "simpler" cell imaging assays, such as measuring size of spheroids or organoids. The challenges of developing information-rich and disease-relevant imaging-based phenotypic HTS assays that are amenable in 3D tissue models have been discussed above and in the literature [99]. As the technical challenges of using fluorescence imaging-based assays in 3D tissue settings are being addressed [78], one should not forget to consider alternative assays that might be still disease-relevant and amenable to HTS. That includes label-free morphology-based imaging, which coupled to machine learning computational approaches has been used to score for cell differentiation phenotypes in 2D [100–103]. Measuring different morphological features of disease iPSC-derived 3D optic cups using bright-field may enable drug screening for a relatively large collection of compounds, which could provide a smaller set of compounds to study in more complex, information-rich assays. In addition, sampling tissue supernatant for disease-relevant markers such as cytokines can also provide HTS-amenable assay readouts, which could be used as a primary screen, again followed by testing a limited number of hits in information-rich, lower throughput assays. In this regard, liquid chromatography-mass spectrometry (LC-MS) and matrix-assisted laser desorption/ionization-mass spectrometry (MALDI-MS) technologies are now available with higher throughput that would enable the "secretome" analysis of supernatant content of normal versus disease 3D tissue models, and perhaps leverage any differences as a HTS readout. It is worth mentioning that, albeit it is a low-throughput read out, MALDI-MS-based imaging of in-tissue sections is becoming a very powerful tool to image biomolecules for which antibodies are not readily available, including lipids [104–106]. As other techniques such as 3D bioprinting are being used to develop layered systems with relevant monolayers (e.g., RPE), measurements such as TEER [107] will enable additional readouts which could be done in at least 96-well microplate format, which should enable focused screens and hit validation. Finally, there are technologies that enable gene expression analysis in HTS settings, including the QuantiGene Plex Gene Expression Assay, the Nanostring technology and the Fluidigm's integrated fluidic circuits' automated PCR technology. An example of the application of one of these technologies to monitor iPSC differentiation to RPE in 96- and 384-well microplate format was reporter by Ferrer et al. [108].

7.5.4   Machine Learning

It is now clear that to maximize the outcome from drug screens using 3D tissue models, the application of machine learning approaches to both cheminformatics [109] and high-content cell imaging assays will be critical [110–113]. The application of machine learning to cheminformatics should allow for the screening of carefully chosen, focused diversity collections to create models of which structural features in compounds produce the desired activity in the assays, and rounds of compound selection based on the prediction models generated, and testing should be able to generate new leads for chemistry development. In addition, screening of annotated libraries coupled to cheminformatics and genomics information on a system should enable the identification of new targets and pathways involved in a process [114]. An extensive amount of work is being done to apply machine learning approaches to score and predict control versus disease phenotypes from cell imaging assays, including work on photoreceptor outer segment formation [115]. Most of this work has so far been done in 2D systems, and the hope is that with the development of techniques that allow HT cell imaging in 3D, those will be applied to this more complex systems.

References

1. Scannell JW, Blanckley A, Boldon H, Warrington B (2012) Diagnosing the decline in pharmaceutical R&D efficiency. Nat Rev Drug Discov 11(3):191–200
2. Hay M, Thomas DW, Craighead JL, Economides C, Rosenthal J (2014) Clinical development success rates for investigational drugs. Nat Biotechnol 32(1):40–51
3. BIO BaA (2016) Clinical development success rates 2006-2015. BIO, Washington
4. Harrison RK (2016) Phase II and phase III failures: 2013-2015. Nat Rev Drug Discov 15(12):817–818
5. Repetto G, del Peso A, Zurita JL (2008) Neutral red uptake assay for the estimation of cell viability/cytotoxicity. Nat Protoc 3(7):1125–1131
6. Trapnell C et al (2010) Transcript assembly and quantification by RNA-Seq reveals unannotated transcripts and isoform switching during cell differentiation. Nat Biotechnol 28(5):511–515
7. Potter H (2003) Transfection by electroporation. Curr Protoc Mol Biol Chapter 9:Unit 9 3
8. Connor KM et al (2009) Quantification of oxygen-induced retinopathy in the mouse: a model of vessel loss, vessel regrowth and pathological angiogenesis. Nat Protoc 4(11):1565–1573
9. McAvoy JW, Chamberlain CG, de Iongh RU, Hales AM, Lovicu FJ (1999) Lens development. Eye (Lond) 13(Pt 3b):425–437
10. Ferrara N, Hillan KJ, Gerber HP, Novotny W (2004) Discovery and development of bevacizumab, an anti-VEGF antibody for treating cancer. Nat Rev Drug Discov 3(5):391–400
11. Rongvaux A et al (2014) Development and function of human innate immune cells in a humanized mouse model. Nat Biotechnol 32(4):364–372
12. Chiu JJ, Chien S (2011) Effects of disturbed flow on vascular endothelium: pathophysiological basis and clinical perspectives. Physiol Rev 91(1):327–387
13. Li YS, Haga JH, Chien S (2005) Molecular basis of the effects of shear stress on vascular endothelial cells. J Biomech 38(10):1949–1971
14. Huh D et al (2010) Reconstituting organ-level lung functions on a chip. Science 328(5986):1662–1668
15. Huh D, Torisawa YS, Hamilton GA, Kim HJ, Ingber DE (2012) Microengineered physiological biomimicry: organs-on-chips. Lab Chip 12(12):2156–2164
16. Newman AC, Nakatsu MN, Chou W, Gershon PD, Hughes CC (2011) The requirement for fibroblasts in angiogenesis: fibroblast-derived matrix proteins are essential for endothelial cell lumen formation. Mol Biol Cell 22(20):3791–3800
17. Chen JX, Stinnett A (2008) Disruption of Ang-1/Tie-2 signaling contributes to the impaired myocardial vascular maturation and angiogenesis in type II diabetic mice. Arterioscler Thromb Vasc Biol 28(9):1606–1613

18. Wakui S et al (2006) Localization of Ang-1, -2, Tie-2, and VEGF expression at endothelial-pericyte interdigitation in rat angiogenesis. Lab Invest 86(11):1172–1184
19. Moya ML, Alonzo LF, George SC (2014) Microfluidic device to culture 3D in vitro human capillary networks. Methods Mol Biol 1202:21–27
20. Wang X, Phan DTT, George SC, Hughes CCW, Lee AP (2017) 3D anastomosed microvascular network model with living capillary networks and endothelial cell-lined microfluidic channels. Methods Mol Biol 1612:325–344
21. Jeon JS et al (2015) Human 3D vascularized organotypic microfluidic assays to study breast cancer cell extravasation. Proc Natl Acad Sci U S A 112(1):214–219
22. Kim S, Lee H, Chung M, Jeon NL (2013) Engineering of functional, perfusable 3D microvascular networks on a chip. Lab Chip 13(8):1489–1500
23. Chung M et al (2018) Wet-AMD on a chip: modeling outer blood-retinal barrier in vitro. Adv Healthc Mater 7(2)
24. Becerra SP et al (2004) Pigment epithelium-derived factor in the monkey retinal pigment epithelium and interphotoreceptor matrix: apical secretion and distribution. Exp Eye Res 78(2):223–234
25. Saint-Geniez M, Kurihara T, Sekiyama E, Maldonado AE, D'Amore PA (2009) An essential role for RPE-derived soluble VEGF in the maintenance of the choriocapillaris. Proc Natl Acad Sci U S A 106(44):18751–18756
26. Sonoda S et al (2009) Attainment of polarity promotes growth factor secretion by retinal pigment epithelial cells: relevance to age-related macular degeneration. Aging 2(1):28–42
27. Bailey TA et al (2004) Oxidative stress affects the junctional integrity of retinal pigment epithelial cells. Invest Ophthalmol Vis Sci 45(2):675–684
28. Hamilton RD, Foss AJ, Leach L (2007) Establishment of a human in vitro model of the outer blood-retinal barrier. J Anat 211(6):707–716
29. Song MJ, Quinn R, Dejene R, Bharti K (2017) 3D tissue engineered RPE/"choroid" to identify mechanism of AMD-disease initiation and progression. Assoc Res Vis Ophthalmol 58(8):3760–3760
30. Song MJ, Bharti K (2016) Looking into the future: using induced pluripotent stem cells to build two and three dimensional ocular tissue for cell therapy and disease modeling. Brain Res 1638(Pt A):2–14
31. Hampton C et al (2018) Hypoxia of retina pigment epithelium induces type 1 CNV-like morphology within 3D engineered iPSC-RPE/"Choroid" tissues. Invest Ophthalmol Vis Sci 59(9):3272–3272
32. Hotaling NA et al (2016) Nanofiber scaffold-based tissue-engineered retinal pigment epithelium to treat degenerative eye diseases. J Ocul Pharmacol Ther 32(5):272–285
33. Ablonczy Z, Crosson CE (2007) VEGF modulation of retinal pigment epithelium resistance. Exp Eye Res 85(6):762–771
34. Curcio CA, Johnson M (2012) Structure, function, and pathology of Bruch's membrane. Elastic:465–481
35. Baba T et al (2009) Maturation of the fetal human choriocapillaris. Invest Ophthalmol Vis Sci 50(7):3503–3511
36. Takahashi K et al (2007) Induction of pluripotent stem cells from adult human fibroblasts by defined factors. Cell 131(5):861–872
37. Reh TA (2017) The development of the retina. Ryan's Retina. Elsevier, Amsterdam
38. Zuber ME, Gestri G, Viczian AS, Barsacchi G, Harris WA (2003) Specification of the vertebrate eye by a network of eye field transcription factors. Development 130(21):5155–5167
39. Lamba DA, Karl MO, Ware CB, Reh TA (2006) Efficient generation of retinal progenitor cells from human embryonic stem cells. Proc Natl Acad Sci U S A 103(34):12769–12774
40. Osakada F et al (2008) Toward the generation of rod and cone photoreceptors from mouse, monkey and human embryonic stem cells. Nat Biotechnol 26(2):215–224
41. Mellough CB, Sernagor E, Moreno-Gimeno I, Steel DH, Lako M (2012) Efficient stage-specific differentiation of human pluripotent stem cells toward retinal photoreceptor cells. Stem Cells 30(4):673–686

42. Hunt NC et al (2017) 3D culture of human pluripotent stem cells in RGD-alginate hydrogel improves retinal tissue development. Acta Biomater 49:329–343

43. Eiraku M et al (2011) Self-organizing optic-cup morphogenesis in three-dimensional culture. Nature 472(7341):51–56

44. Nakano T et al (2012) Self-formation of optic cups and storable stratified neural retina from human ESCs. Cell Stem Cell 10(6):771–785

45. Kuwahara A et al (2015) Generation of a ciliary margin-like stem cell niche from self-organizing human retinal tissue. Nat Commun 6:6286

46. Wahlin KJ et al (2017) Photoreceptor outer segment-like structures in long-term 3D retinas from human pluripotent stem cells. Sci Rep 7(1):766

47. Volkner M et al (2016) Retinal organoids from pluripotent stem cells efficiently recapitulate retinogenesis. Stem Cell Rep 6(4):525–538

48. Gonzalez-Cordero A et al (2013) Photoreceptor precursors derived from three-dimensional embryonic stem cell cultures integrate and mature within adult degenerate retina. Nat Biotechnol 31(8):741–747

49. Meyer JS et al (2009) Modeling early retinal development with human embryonic and induced pluripotent stem cells. Proc Natl Acad Sci U S A 106(39):16698–16703

50. Meyer JS et al (2011) Optic vesicle-like structures derived from human pluripotent stem cells facilitate a customized approach to retinal disease treatment. Stem Cells 29(8):1206–1218

51. Zhong X et al (2014) Generation of three-dimensional retinal tissue with functional photoreceptors from human iPSCs. Nat Commun 5:4047

52. Luo Z et al (2018) An optimized system for effective derivation of three-dimensional retinal tissue via wnt signaling regulation. Stem Cells 36:1709

53. Zhu Y et al (2013) Three-dimensional neuroepithelial culture from human embryonic stem cells and its use for quantitative conversion to retinal pigment epithelium. PLoS One 8(1):e54552

54. Lowe A, Harris R, Bhansali P, Cvekl A, Liu W (2016) Intercellular adhesion-dependent cell survival and ROCK-regulated actomyosin-driven forces mediate self-formation of a retinal organoid. Stem Cell Rep 6(5):743–756

55. Shirai H et al (2016) Transplantation of human embryonic stem cell-derived retinal tissue in two primate models of retinal degeneration. Proc Natl Acad Sci U S A 113(1):E81–E90

56. Mandai M et al (2017) Autologous induced stem-cell-derived retinal cells for macular degeneration. N Engl J Med 376(11):1038–1046

57. Dorrie J, Wellner V, Kampgen E, Schuler G, Schaft N (2006) An improved method for RNA isolation and removal of melanin contamination from melanoma tissue: implications for tumor antigen detection and amplification. J Immunol Methods 313(1-2):119–128

58. Eckhart L, Bach J, Ban J, Tschachler E (2000) Melanin binds reversibly to thermostable DNA polymerase and inhibits its activity. Biochem Biophys Res Commun 271(3):726–730

59. Chung JY et al (2016) A melanin-bleaching methodology for molecular and histopathological analysis of formalin-fixed paraffin-embedded tissue. Lab Invest 96(10):1116–1127

60. Liu CH et al (2013) Melanin bleaching with dilute hydrogen peroxide: a simple and rapid method. Appl Immunohistochem Mol Morphol 21(3):275–279

61. Kim SY, Assawachananont J (2016) A new method to visualize the intact subretina from retinal pigment epithelium to retinal tissue in whole mount of pigmented mouse eyes. Transl Vis Sci Technol 5(1):6

62. Thanos A et al (2012) Evidence for baseline retinal pigment epithelium pathology in the Trp1-Cre mouse. Am J Pathol 180(5):1917–1927

63. Tainaka K, Kuno A, Kubota SI, Murakami T, Ueda HR (2016) Chemical principles in tissue clearing and staining protocols for whole-body cell profiling. Annu Rev Cell Dev Biol 32:713–741

64. Silvestri L, Costantini I, Sacconi L, Pavone FS (2016) Clearing of fixed tissue: a review from a microscopist's perspective. J Biomed Opt 21(8):081205

65. Jacques SL (2013) Optical properties of biological tissues: a review. Phys Med Biol 58(11):R37–R61
66. Spalteholz W (1914) Über das Durchsichtigmachen von menschlichen und tierischen Präparaten und seine theoretischen Bedingungen, nebst Anhang: Über Knochenfärbung. S. Hirzel, Leipzig
67. Erturk A et al (2012) Three-dimensional imaging of solvent-cleared organs using 3DISCO. Nat Protoc 7(11):1983–1995
68. Ke MT, Fujimoto S, Imai T (2013) SeeDB: a simple and morphology-preserving optical clearing agent for neuronal circuit reconstruction. Nat Neurosci 16(8):1154–1161
69. Hama H et al (2011) Scale: a chemical approach for fluorescence imaging and reconstruction of transparent mouse brain. Nat Neurosci 14(11):1481–1488
70. Hama H et al (2015) ScaleS: an optical clearing palette for biological imaging. Nat Neurosci 18(10):1518–1529
71. Susaki EA, Ueda HR (2016) Whole-body and whole-organ clearing and imaging techniques with single-cell resolution: toward organism-level systems biology in mammals. Cell Chem Biol 23(1):137–157
72. Chung K et al (2013) Structural and molecular interrogation of intact biological systems. Nature 497(7449):332–337
73. Yang B et al (2014) Single-cell phenotyping within transparent intact tissue through whole-body clearing. Cell 158(4):945–958
74. Murray E et al (2015) Simple, scalable proteomic imaging for high-dimensional profiling of intact systems. Cell 163(6):1500–1514
75. Zheng H, Rinaman L (2016) Simplified CLARITY for visualizing immunofluorescence labeling in the developing rat brain. Brain Struct Funct 221(4):2375–2383
76. Phillips J et al (2016) Development of passive CLARITY and immunofluorescent labelling of multiple proteins in human cerebellum: understanding mechanisms of neurodegeneration in mitochondrial disease. Sci Rep 6:26013
77. Kuwajima T et al (2013) ClearT: a detergent- and solvent-free clearing method for neuronal and non-neuronal tissue. Development 140(6):1364–1368
78. Boutin ME et al (2018) A high-throughput imaging and nuclear segmentation analysis protocol for cleared 3D culture models. Sci Rep 8(1):11135
79. Grist SM, Nasseri SS, Poon T, Roskelley C, Cheung KC (2016) On-chip clearing of arrays of 3-D cell cultures and micro-tissues. Biomicrofluidics 10(4):044107
80. Silva Santisteban T, Rabajania O, Kalinina I, Robinson S, Meier M (2017) Rapid spheroid clearing on a microfluidic chip. Lab Chip 18(1):153–161
81. Rajasekaran B, Uriu K, Valentin G, Tinevez JY, Oates AC (2016) Object segmentation and ground truth in 3D embryonic imaging. PLoS One 11(6):e0150853
82. Li L, Zhou Q, Voss TC, Quick KL, LaBarbera DV (2016) High-throughput imaging: focusing in on drug discovery in 3D. Methods 96:97–102
83. Schmitz A, Fischer SC, Mattheyer C, Pampaloni F, Stelzer EH (2017) Multiscale image analysis reveals structural heterogeneity of the cell microenvironment in homotypic spheroids. Sci Rep 7:43693
84. Jones TR et al (2009) Scoring diverse cellular morphologies in image-based screens with iterative feedback and machine learning. Proc Natl Acad Sci U S A 106(6):1826–1831
85. Inglese J et al (2007) High-throughput screening assays for the identification of chemical probes. Nat Chem Biol 3(8):466–479
86. Ko HC, Gelb BD (2014) Concise review: drug discovery in the age of the induced pluripotent stem cell. Stem Cells Transl Med 3(4):500–509
87. Haasen D et al (2017) How phenotypic screening influenced drug discovery: lessons from five years of practice. Assay Drug Dev Technol 15(6):239–246
88. Ursu A, Scholer HR, Waldmann H (2017) Small-molecule phenotypic screening with stem cells. Nat Chem Biol 13(6):560–563

89. Smith K et al (2018) Phenotypic image analysis software tools for exploring and understanding big image data from cell-based assays. Cell Syst 6(6):636–653

90. Kaewkhaw R et al (2016) Treatment paradigms for retinal and macular diseases using 3-D retina cultures derived from human reporter pluripotent stem cell lines. Invest Ophthalmol Vis Sci 57(5):ORSFl1

91. Fuller JA et al (2014) A high content screening approach to identify molecules neuroprotective for photoreceptor cells. Adv Exp Med Biol 801:773–781

92. Chang YC et al (2014) The generation of induced pluripotent stem cells for macular degeneration as a drug screening platform: identification of curcumin as a protective agent for retinal pigment epithelial cells against oxidative stress. Front Aging Neurosci 6:191

93. Ito SI, Onishi A, Takahashi M (2017) Chemically-induced photoreceptor degeneration and protection in mouse iPSC-derived three-dimensional retinal organoids. Stem Cell Res 24:94–101

94. Vergara MN et al (2017) Three-dimensional automated reporter quantification (3D-ARQ) technology enables quantitative screening in retinal organoids. Development 144(20):3698–3705

95. Parfitt DA et al (2016) Identification and correction of mechanisms underlying inherited blindness in human iPSC-derived optic cups. Cell Stem Cell 18(6):769–781

96. Zhou T et al (2017) High-content screening in hPSC-neural progenitors identifies drug candidates that inhibit zika virus infection in fetal-like organoids and adult brain. Cell Stem Cell 21(2):274–283. e275

97. Mathews Griner LA et al (2016) Large-scale pharmacological profiling of 3D tumor models of cancer cells. Cell Death Dis 7(12):e2492

98. Hou S et al (2018) Advanced development of primary pancreatic organoid tumor models for high-throughput phenotypic drug screening. SLAS Discov 23(6):574–584

99. Carragher N et al (2018) Concerns, challenges and promises of high-content analysis of 3D cellular models. Nat Rev Drug Discov 17:606

100. Fujitani M et al (2017) Morphology-based non-invasive quantitative prediction of the differentiation status of neural stem cells. J Biosci Bioeng 124(3):351–358

101. Kobayashi H et al (2017) Label-free detection of cellular drug responses by high-throughput bright-field imaging and machine learning. Sci Rep 7(1):12454

102. Matsuoka F et al (2013) Morphology-based prediction of osteogenic differentiation potential of human mesenchymal stem cells. PLoS One 8(2):e55082

103. Sasaki H et al (2014) Label-free morphology-based prediction of multiple differentiation potentials of human mesenchymal stem cells for early evaluation of intact cells. PLoS One 9(4):e93952

104. Anderson DM et al (2014) High resolution MALDI imaging mass spectrometry of retinal tissue lipids. J Am Soc Mass Spectrom 25(8):1394–1403

105. Deutskens F, Yang J, Caprioli RM (2011) High spatial resolution imaging mass spectrometry and classical histology on a single tissue section. J Mass Spectrom 46(6):568–571

106. Seeley EH, Schwamborn K, Caprioli RM (2011) Imaging of intact tissue sections: moving beyond the microscope. J Biol Chem 286(29):25459–25466

107. Srinivasan B et al (2015) TEER measurement techniques for in vitro barrier model systems. J Lab Autom 20(2):107–126

108. Ferrer M et al (2014) A multiplex high-throughput gene expression assay to simultaneously detect disease and functional markers in induced pluripotent stem cell-derived retinal pigment epithelium. Stem Cells Transl Med 3(8):911–922

109. Lo YC, Rensi SE, Torng W, Altman RB (2018) Machine learning in chemoinformatics and drug discovery. Drug Discov Today 23:1538

110. Horvath P, Wild T, Kutay U, Csucs G (2011) Machine learning improves the precision and robustness of high-content screens: using nonlinear multiparametric methods to analyze screening results. J Biomol Screen 16(9):1059–1067

111. O'Duibhir E et al (2018) Machine learning enables live label-free phenotypic screening in three dimensions. Assay Drug Dev Technol 16(1):51–63

112. Piccinini F et al (2017) Advanced cell classifier: user-friendly machine-learning-based software for discovering phenotypes in high-content imaging data. Cell Syst 4(6):651–655. e655
113. Smith K, Horvath P (2014) Active learning strategies for phenotypic profiling of high-content screens. J Biomol Screen 19(5):685–695
114. Strang BL et al (2018) Identification of lead anti-human cytomegalovirus compounds targeting MAP4K4 via machine learning analysis of kinase inhibitor screening data. PLoS One 13(7):e0201321
115. Fuller JA, Berlinicke CA, Inglese J, Zack DJ (2016) Use of a machine learning-based high content analysis approach to identify photoreceptor neurite promoting molecules. Adv Exp Med Biol 854:597–603

Index

A

Adaptive immune system, 103
Adeno-associated viral (AAV), 10
Adult-onset vitelliform macular dystrophy
(AVMD), 7
Age-related macular degeneration (AMD),
2–4, 38, 58, 59, 142, 172
Alcon Constellation®, 158
Alcon NGENUITY® 3D Visualization
System, 158
Allogeneic transplantation, 108
Anesthesia
animal positioning, 153
eye-specific preparations, 153
macaques/non-human primates, 152
pig, 152
rabbit
eye-specific preparations, 151, 152
head positioning, 151
pain management, 151
premedication, 151
rectal thermometer, 151
small rodents (mice and rats), 150
Angiopoietin-1 (ANG-1), 173
Anterior Chamber-Associated Immune
Deviation (ACAID), 82
Apolipoprotein E (*APOE*), 8
Assay systems, 183
Autosomal dominant radial drusen
(ADRD), 13
Autosomal dominant
vitreoretinochoroidopathy
(ADVIRC), 7
Autosomal recessive bestrophinopathy
(ARB), 7

B

Balance salt solution (BSS), 155
Basic fibroblast growth factor (bFGF), 66
Basiliximab, 109
Best disease (BD), 7, 38
Best vitelliform macular dystrophy
(BVMD), 16
Bioink, 174, 175
Bioprinting, 174
Bipolar cells, 124
Bleb retinal detachment (bRD), 157, 162, 163
Blindness, 57
Blood–brain barrier (BBB), 100
Bone marrow mesenchymal stem cells
(MSCs), 67, 68
Bone morphogenic protein (BMP), 178
Brain-derived neurotrophic factor (BDNF), 65

C

Calcium- and magnesium-containing
Hank's balanced salt solution
(CM-HBSS), 156
Cell rejection
allogenic cells, 107, 108
circumvent, 108, 109
prevention, 108
Cell replacement, 100, 106, 107, 132
CellTiterGlo, 185
Cell transplantation, 63, 64, 99, 100
Cellular replacement strategies, 123
Cellular retinaldehyde-binding protein
(*CRALBP*), 8
Cellular systems, 172
Cell viability assay, 185

© Springer Nature Switzerland AG 2019
K. Bharti (ed.), *Pluripotent Stem Cells in Eye Disease Therapy*,
Advances in Experimental Medicine and Biology 1186,
https://doi.org/10.1007/978-3-030-28471-8

Central nervous system (CNS), 100, 125
Chemical biology studies, 184
Chetomin, 70
Chondroitin sulfate proteoglycan
　　　(CSPG), 105
Choriocapillaris, 175, 177
Choroid, 174
Choroidal neovascularization (CNV), 4, 63, 174
Choroideremia, 38
Ciliary neurotrophic factor (CNTF), 66
Complement system, 103, 104
Connective tissue growth factor
　　　(CTGF), 41
CRISPR/Cas technology, 143
CRXp-GFP reporter, 184
Cyclosporine, 107, 146
Cyclosporine A, 62
Cynomolgus (*Macaca fascicularis*), 146

D
Dendritic cells (DC), 107
3,3′-Diaminobenzidine (DAB), 181
Differentiated RPE cells
　　　functionality, 76
　　　purity, 75
　　　safety, 73–75
Disk diameter (DD), 162
Dominant optic atrophy (DOA), 129
Doxycycline, 148
Doyne's honeycomb retinal dystrophy/
　　　Malattia Leventinesse
　　　(DHRD/ML), 4–6
Drug development, 123
Drusen, 4, 9
Dual leucine zipper kinase (DLK)
　　　pathway, 130

E
EGF-containing fibulin extracellular matrix
　　　protein 1 (EFEMP1), 3
Electric cell-substrate impedance sensing
　　　(ECIS), 13
Electroretinogram (ERG), 61
Embryoid body (EB), 70
Embryonic development, 122
Embryonic stem cells (ESCs), 34, 178
Endoplasmic reticulum (ER), 7
Epidermal growth factor (EGF), 41
Epithelial–mesenchymal transition (EMT), 40,
　　　41, 43
Extracellular matrix (ECM), 3, 39, 132

F
Fenestrated endothelial cell linked protein
　　　(FELS), 177
Fenestration marker, 177
Fibroblast growth factor (FGF), 178
Fibroblasts, 173
Fluid–air exchange (FAX), 157
Fluorescence imaging-based assays, 187
Fluorescence reporter, 184
4-hydroxytamoxifen (4-OHT), 185

G
Genetic disease models, 130
Genetic engineering, 143
Glasgow Minimum Essential Medium
　　　(GMEM), 72
Glial-derived neurotrophic factor (GDNF), 66
Global Haplotype Cell Banking Initiative,
　　　84, 85
Glucocorticoids, 109
Green fluorescent protein (GFP), 67
Gyrate atrophy, 16, 17

H
Hematoxylin and eosin (H&E), 181
Hepatocyte growth factor (HGF), 41
High content screening (HCS), 184
High-performance liquid chromatography
　　　(HPLC), 37
High-throughput drug screening (HTS), 34,
　　　183–185
hiPSC-based disease modeling, 18–22
Human amniotic membrane (hAM), 77
Human central nervous system stem cells
　　　(HuCNS-SCs), 67
Human cortical progenitor cells, 66
Human embryonic stem cells (hESCs), 69
Human hematopoietic stem cells (HSCs), 173
Human-induced pluripotent stem cells
　　　(hiPSCs), 69
　　　AMD, 12, 14
　　　CNV, 12
　　　ECM, 13
　　　L-ORD, 13
　　　PGC-1α, 15
　　　ROS, 13
　　　RP, 15, 16
　　　RPE characteristics, 12
　　　RPE-related disorders, 16–18
　　　sources, 12
　　　tissue-on-chip approach, 13

Human pluripotent stem cell (hPSC), *see*
    Retinal ganglion cells (RGCs)
Human umbilical tissue-derived cells
    (hUTCs), 68

I
Immune modulation
    factors, 110
    synthetic, 110, 111
Immune privilege of eye, 82–84
Immune system development
    adaptive, 103
    BBB, 100
    BRB, 100
    complement system, 103, 104
    innate, 101, 103
    protein-based complement system, 100
    TGF-β, 100
Immunosuppression
    monkey, 150
    mouse, 146
    pig, 148, 149
    rabbit, 148
    rat, 146
Immunosuppressive treatment, 84, 85
InCuyte Zoom, 186
Induced pluripotent stem cells (iPSCs), 69,
    107, 173
Innate immune system, 101, 103
Inner limiting membrane (ILM), 156
Insulin-like growth factor 1 (IGF-1), 110
Interferon beta (IFN-β), 111
Intraocular pressure (IOP), 155, 161
Intraoperative optical coherence tomography
    (iOCT), 158, 163
Investigational new drug (IND), 172
In vitro models
    anti-inflammatory agents, 39
    β-catenin signaling, 42
    blood–ocular barriers, 39
    cellular signaling mechanisms, 41
    drug screening, 38
    EMT, 40
    injury, 38
    iPS-RPE, 42
    myofibroblasts, 41
    PDGF, 41
    post-traumatic intraocular fibrotic
        conditions, 38
    PVR, 39, 40, 42–44
    retinal detachment, 40
    TGFβ, 41
    wound healing, 38

iPSC-derived cells, 173
Iris pigment epithelium (IPE), 65, 66

K
Knockout serum replacement (KSR), 64, 72

L
Late-onset macular degeneration (L-ORMD), 6
Late-onset retinal degeneration (L-ORD), 4,
    6, 7
LCHAD deficiency, 38
Leber congenital amaurosis (LCA), 2, 16, 17,
    58, 186
Lecithin retinol acyltransferase (LRAT), 8, 57
Leucine zipper kinase (LZK) pathway, 130
Liquid chromatography-mass spectrometry
    (LC-MS), 187
Low-density lipoprotein (LDL), 8
Lymphocytes, 103

M
Macrophages, 101, 110
Macula, 143
Matrix-assisted laser desorption/ionization-
    mass spectrometry (MALDI-MS)
    technologies, 187
Matrix-metalloproteases (MMPs), 5
Mechanical traction, 156
Mechanism interrogation plate (MIPE), 184
Membrane attack complex (MAC), 9, 106
Membrane-type frizzled-related protein
    (MFRP), 8
Mendelian retinal disorders, 2
Mer tyrosine kinase receptor (MERTK), 8
Mesencephalic astrocyte-derived neurotrophic
    factor (MANF), 110
Methyl thiazol tetrazolium (MTT) assay, 185
MicroDose™ injection kit, 159
Microglia, 101, 102
Microglia/macrophage activation
    anti-inflammatory cytokines, 105
    CD40, 106
    CSPG, 105
    gliosis, 106
    IL-33, 106
    inflammasome, 106
    intracellular miRNA-155 expression, 104
    MAC, 106
    M2 transition process, 105
    neurotrophic factors, 105
    oligodendrogenesis, 105

Microglia/macrophage activation (*cont.*)
  proinflammatory and cytotoxic factors, 105
  remyelination, 105
  retinal disorders, 102
  retinal inflammation process, 104
  VEGF, 105
Microphthalmia-associated transcription factor
      (MITF), 178
MicroRNAs (miRNAs), 43
Microvitreoretinal (MVR), 155
Minimal-invasive procedure, 164
Minocycline, 110, 148
Monkey, 150
Monkey surgery
  implantation, 162, 163
  instrument and machine preparations, 160
  preparing implantation instrumentation, 162
  vitrectomy, 160, 161
  wound closure, 163
Monocyte chemoattractant protein 1 (MCP1), 9
Mouse, 146
Mouse embryonic fibroblast (MEF), 72
Müller glia, 103, 106, 124
Multipotent stem cells, 34

N
Nanostring technology, 187
Nerve fiber layer, 124
Neural retina leucine zipper (NRL), 179
Neurodegeneration, 123
Neuronal cell types, 125
Neuroprotective strategies, 130
Neurosensory retina, 174
Neurotransmitter, 124
Neurotrophins, 69
Neutrophils, 101
Non-human primate, 162
Nonsense mediated decay (NMD), 17
Normal tension glaucoma (NTG), 128

O
Ocular tissue models
  ESC/iPSC-derived cells, 185
  HTS, 183–185
Ophthalmology, 174
Optic atrophy, 129
Optic nerve, 124
Optic neuropathies, 122, 123, 132
Optineurin (OPTN), 128
Optokinetic paradigm, 66
Outer blood–retina barrier, *see* 3D engineering

P
Patient-specific cells, 130
Perfluorcarbon liquids (PFC), 157
Pericytes, 173
Peripheral layers, 124
Pharmacological approaches, 123
Photodynamic therapy, 60
Photoreceptor outer segments (POS),
      3, 7, 16
Photoreceptors (PR), 123, 124, 142,
      174, 175
Photoreceptor transplantation
  graft preparation, 164
  transplant delivery strategies, 164
Phototransduction, 56
Pig surgery
  bleb retinal detachment, 160
  implantation, 158
  instrument and equipment preparation, 157
  nasal sclerotomy, 160
  retinal detachment, 158
  retinotomy, 160
Pigment epithelium-derived factor
      (PEDF), 3, 174
Pigs, 149, 152
  domestic, 145
  mini-pigs, 145
  Yucatan, 146
Platelet-derived growth factor (PDGF), 41
Pluripotent stem cell (PSC), 34
Porous polyester terephthalate (PET), 154
Posterior vitreous detachment (PVD),
      155, 161
Post-traumatic ocular fibrosis, 44
Preclinical study, 143, 164
Proliferative vitreoretinopathy (PVR)
  advantages, 43
  categories, 39
  cellular basis, 40
  compounds, 42
  ECM, 39
  epiretinal/subretinal membranes, 41
  etiology, 40
  iPS-RPE model, 44
  medical treatments, 39
  ocular trauma, 39
  pathogenesis, 39–42
  pathology, 42
  progression, 39
  retinal detachments, 40
  surgical techniques, 39
  vitreoretinal surgery, 39
Punctate inner choroidopathy, 39

Q
QuantiGene Plex Gene Expression Assay, 187

R
Rabbit surgery
    implantation, 156, 157
    instrument preparation, 155
    loading implantation instrument, 156
    vitrectomy, 155, 156
    wound closure, 157
Rabbits, 144, 145, 147, 148
Rapamicin, 149
Rat, 146
Reactive oxygen species (ROS), 13, 14, 185
Reprogramming techniques, 35
Resight®, 158
Retinal cell types, 123
Retinal degenerative diseases (RDDs)
    AMD, 2–4, 8
    BD, 7, 10
    cell types, 2
    complement system, 9
    DHRD/ML, 5, 6
    electron lucent particles, 8
    genetic diseases, 8
    knock-in mice model, 10
    light absorption, 2
    L-ORD, 6, 7, 10
    MERTK, 11
    non-human primates, 11
    oxidative stress, 8, 9
    POS, 3
    retinoids, 3
    RP, 7, 8
    RPE functions, 2
    SFD, 5
    Y204H mutation, 9
Retinal development, 122
Retinal ganglion cells (RGCs)
    applications, 122, 123
    BRN3 expression, 127
    cell types, 124
    degeneration and pathogenesis, 122
    future applications
        apoptotic and autophagy pathways, 133
        cell replacement strategies, 132
        CRIPSR engineering, 133
        development and disease pathologies, 130
        early disease progression, 132
        peripheral layers, 131
        retinal organoids, 131
        spatial lamination, 131

    subtypes, 132
    temporal development, 131
    genetic markers, 125–127
    human-derived cells, 127
    molecular signatures, 127
    optic neuropathies, 122
    translational applications
        animal models, 128
        BDNF and GDNF, 129
        CRISPR gene editing, 130
        disease-causing mutations, 128
        human degenerative diseases, 128
        neural and retinal cells, 128
        optic neuropathies, 128, 129
        OPTN, 129
        TBK1, 128
Retinal neurospheres (RNS), 68
Retinal organoids, 124, 131
Retinal pigment epithelium (RPE), 174
    adult and fetal, 78
    adult eye, 63, 64
    animal preparation, 154
    bone marrow MSCs, 67, 68
    cell sheet implantation, 155
    cell sheet preparation, 154
    cell suspension injection, 155
    cell suspension preparation, 154
    dystrophies, 57
    fetal brain-derived neural progenitors, 66, 67
    fetal-derived cells, 61, 62
    hiPSCs/hESCs, 78–80
    injection strategy, 76–78
    IPE, 65, 66
    logistic and management advice, 144
    non-human primates, 146
    photoreceptors, 56
    pigs, 145
    pluripotent stem cells
        clinical translation, 71
        continuous adherent culture method, 69
        culture medium, 72
        EBs, 70, 72
        embryoid body method, 69
        exogenous molecules, 71
        in vitro fertilization, 69
        neuroectodermal differentiation, 70
        neuroepithelium structure, 71
        pigmented regions, 70
        RC-09, 72
        TGF-ß pathway, 70
        transcription factors, 69
        types, 69
        WNT pathway, 70

Retinal pigment epithelium (RPE) (*cont.*)
   quality controls (*see* Differentiated
      RPE cells)
   rabbits, 144, 145
   retinopathies (*see* Retinopathies)
   RNS, 68
   Schwann cells, 66
   small rodents, 144
   stem cell trials, 80
   umbilical-cord stem cells, 68
Retinal pigment epithelium derived from
      induced pluripotent stem cells
      (iPS-RPE), 35, 36, 38
Retinitis pigmentosa (RP), 2, 7, 8, 57, 58, 186
Retinitis pigmentosa 50 (RP50), 7
Retinoid X receptor (RXR) gamma, 179
Retinol, 3
Retinopathies
   AMD, 58, 59
   global burden evolution and treatments,
      59, 60
   interconnected neurons, 56
   photoreceptors, 56
   RP, 57, 58
   SRS, 56
Retinotomy edges, 163
Retrobulbar injections, 153
Retroviral gene delivery systems, 35
Reverse transcriptase polymerase chain
      reaction (RT-PCR), 66
Rhesus monkeys (*Macaca mulatta*), 146
RNA-binding protein with multiple splicing
      (RBMPS), 127
Royal College of Surgeons (RCS), 144
RPE stem cells (RPESCs), 63, 64

S
Scaffolding device, 132
Schwann cells, 66
Silent information regulator T1 (SIRT1), 15
Sirolimus, 148
Sorsby's fundus dystrophy (SFD), 4, 5
Specific pathogen-free (SPF), 146
Spontaneous mutations, 142
Stargardt's disease, 2, 38
Stargardt's macular dystrophy (SMD), 38, 78
Stem cell sources, 34
Subretinal space (SRS), 56
Subretinal transplantation, 164
Sunitinib, 130
Superoxide dismutase 1 (SOD1), 8

T
Tacrolimus, 149
TANK Binding Kinase 1 (TBK1), 128
Teratogenicity assays, 142
Teratomas, 74
3D automated reporter quantification
      (3D-ARQ) assay, 186
3D engineering
   animal model systems, 172
   bioprinting, 174
   cellular systems, 172
   diencephalon, 177
   disease cellular and molecular
      mechanisms, 172
   engineered systems, 175
   eye and neurosensory retina, 177
   eye-field transcription factors, 178
   fibroblasts and pericytes, 173
   fluidic channel, 173
   HTS readouts and validation, 186, 187
   human iPSC-derived retinal cup, 185, 186
   in vitro cell assays, 172
   in vivo mouse models, 172
   machine learning, 187
   microfluidics, 173
   ophthalmology, 174
   pharmaceutical and biotechnology
      industries, 172
   physiological differences, 173
   pluripotent stem cell-derived optic cup,
      178–180
   segmentation and quantification, 182
   stem cell technology, 177
   tissue clearing, 181, 182
   traditional histology, 180, 181
   validation, 176
   vascular channel, 173
   vascular endothelial cells, 173
TIMP metallopeptidase inhibitor 3 (TIMP3), 3
Transcriptomics analysis, 184
Transepithelial electrical resistance (TEER), 175
Transforming growth factor α (TGFα), 41
Transforming growth factor beta (TGF-β), 41,
      100
Translational read-through-inducing drugs
      (TRIDs), 16
Transmission electrical microscopy
      (TEM), 175
Transscleral injection, 164
Transvitreal injection, 164
Traumatic maculopathy, 39
Tumorigenicity, 74
Tumor spheroids, 187

U
Unfolded protein response (UPR), 6
US food and drug administration (FDA), 144

V
Vascular endothelial cells, 173
Vascular endothelial growth factor (VEGF), 3,
41, 105

Very low-density lipoprotein (VLDL), 8
Visual impairment, 58
Vitrectomy, 155

W
Wnt pathways, 178
World Precision Instruments, 154
Wound healing, 34, 38, 42–44

Printed in the United States
By Bookmasters